Transitions in Domestic Consumption and Family Life in the Modern Middle East

Transitions in Domestic Consumption and Family Life in the Modern Middle East: Houses in Motion

Edited by

Relli Shechter

TRANSITIONS IN DOMESTIC CONSUMPTION AND FAMILY LIFE IN THE
MODERN MIDDLE EAST

First published 2003 by
PALGRAVE MACMILLAN™
175 Fifth Avenue, New York, N.Y. 10010 and
Houndmills, Basingstoke, Hampshire, England RG21 6XS
Companies and representatives throughout the world

PALGRAVE MACMILLAN is the global academic imprint of the Palgrave Macmillan division of St. Martin's Press, LLC and of Palgrave Macmillan Ltd. Macmillan® is a registered trademark in the United States, United Kingdom and other countries. Palgrave is a registered trademark in the European Union and other countries.

ISBN 1–4039–6189–1 hardback

Library of Congress Cataloging-in-Publication Data

Transitions in domestic consumption and family life in the modern Middle East: houses in motion/edited by Relli Shechter.
    p. cm.
    Includes bibliographical references and index.
    ISBN 1–4039–6189–1
    1. Consumption (Economics)—Middle East—History—Congresses.
2. Households—Middle East—History—Congresses. 3. Family—Middle East—History—Congresses. I. Shechter, Relli.

HC415.15.Z9 C68 2003
339.4′7′0956—dc21   2003056463

A catalogue record for this book is available from the British Library.

Design by Newgen Imaging Systems (P) Ltd., Chennai, India.

First edition: November, 2003
10 9 8 7 6 5 4 3 2 1

Printed in the United States of America.

Table of Contents

List of Figures

List of Tables

Acknowledgments

This edited volume started as a 2001 workshop at Ben-Gurion University, Israel, entitled "Considering Consumption, Production, and the Market in the Constitution of Meaning in the Middle East and Beyond." The workshop was sponsored by the Department of Middle East Studies and the Haim Herzog Center of Middle East Research and Diplomacy. A generous financial contribution from the Soref Foundation allowed the workshop to take place.

Tally Katz-Gerro encouraged me to go ahead with this project once the workshop was over. David Pervin at Palgrave Publishers was very encouraging at all stages of production. Ayşe Buğra and Nancy Micklewright improved on earlier drafts of the introduction. Michal Pinkas meticulously prepared the bibliography.

Notes on the Contributors

Ayşe Buğra is Professor of Political Economy at Boğazici University, Istanbul. Her fields of specialization are development economics, comparative political economy, history and methodology of economics. She is the author of numerous articles and three books. Her latest book is *Islam in Economic Organizations* (TESEV/Friedrich Ebert Foundation, 1999).

Tania Forte completed her doctorate in cultural anthropology at the University of Chicago, and is currently a Lady Davis postdoctoral fellow at the Hebrew University in Jerusalem. She is also a part-time lecturer in the Department of Communications Studies at Ben-Gurion University.

Tally Katz-Gerro is Assistant Professor in the Department of Sociology and Anthropology at the University of Haifa. She writes on the sociology of consumption, sociology of culture, and inequality and social stratification. A list of her recent publications can be found at http://soc.haifa.ac.il/~tkatz

Sonja Laden is lecturing in the Translation Studies Program at the Porter School of Cultural Studies, Tel Aviv University, and is a Research Associate at the Porter Institute for Poetics and Semiotics. Her research interests include comparative analyses of the magazine-form in promoting consumer culture in Israel and South Africa. She has published a number of articles on both these topics, and is the editor (with Louise Bethlehem and Leon de Kock) of *South Africa in the Global Imaginary* (Kwela, forthcoming).

Nancy Micklewright is Program Officer at the Getty Grant Program in Los Angeles. Her research interests include the art and architecture of the late Ottoman Empire, particularly Istanbul and Cairo. Recently she has been working on photography in the Ottoman context. She has published numerous articles on Ottoman painting and photography, women's dress, and gender. Her most recent book is *Annie Lady Brassey, A Victorian Traveler in the Middle East* (Ashgate, 2003).

Lisa Pollard is Associate Professor of History at the University of North Carolina at Wilmington. She is the author of *Nurturing the Nation: The Family Politics of Modernizing, Colonizing and Liberating Egypt (1805–1922)* (The University of California Press, forthcoming) and coeditor (with Lynne Haney) of *Families of a New World: Gender, Politics and State-Building in a Global Context* (Routledge, 2003).

Mona Russell is lecturing in History at MIT. She has published articles on advertising, education, and the use of textbooks as a source for history. She is the author *Creating the New Woman: Consumerism, Education, and National Identity in Egypt, 1863–1922* (Palgrave, forthcoming).

Relli Shechter is Assistant Professor at the Department of Middle East Studies, Ben-Gurion University. He has published articles on consumption and production of tobacco and cigarette in Egypt, advertising, and markets in the Middle East.

Introduction

Relli Shechter

On first sight, the family is the core of intimate life away from society, where the home constitutes the physical and symbolic private sphere. Conceived as such, domestic life in the Middle East and elsewhere was studied until recently in isolation from broader economic, political, and cultural transformations. This volume argues for a different positioning of private and public through the exploration of various venues of actual and symbolic interactions between the two, and the impact of such interactions on domestic consumption. It brings political economy, hence the issues pertaining to the state and the market, and ideology in the form of modernity and nationalism, to bear on family life and the shape of its closest environment: homes and their contents.

The volume focuses on three countries, Egypt, Israel, and Turkey (formerly all included in the Ottoman Empire), in the period from the mid-nineteenth to the early twenty-first century. The chapters are novel ventures into the architecture of the house, its furniture, electric appliances and decorations, and their representation in books, magazines, and the visual arts, especially photography, which attempt to unravel the relations between family, home, and society at large. With chapters coming from various scholarly traditions—anthropology, history and art history, literary criticism, political economy, and sociology—the interdisciplinary nature of the volume constitutes a spectrum of possible approaches to the topic rather than a unified conclusion. But all chapters share a focus on domestic consumption as an important key to understanding the impact of larger changes on our intimate lives.

The volume is situated within the context of contemporary currents of analysis in the scholarship on the Middle East, where recent studies on the family and domestic life have attempted to transcend the previously perceived gap between family and society. Scholarship in various fields has begun to deliver a more nuanced account on the interplay between wider economic, social, and political changes on the one hand and family transitions on the other. Although historical research, with few exceptions,[1] has been slow to venture into studies that mainly focus on the family, Ottoman history has begun to assign more agency to elite households in shaping politics and society.[2] Historians, as well as other researchers of gender, have contributed to covering at least partially previous gaps in scholarship, by evincing a new interest in women's, and to a lesser extent men's, everyday life, their legal and religious status, and their social roles in studies that politicize family relations.[3] The family has also become the locus of attention in demographic research on the changing size of the family and family planning, which have made a huge impact on the structure of societies in developing countries.[4] Even more so, the household and its survival strategies have been studied in an attempt to measure and reflect on the actual impact of economic transformations on individuals in ways that differ from the glossy aggregate statistical measurements of standard econometrics.[5]

The present volume introduces a new perspective in the study of the family. It demonstrates that through studies of domestic consumption the most searching investigations of translations of wider shifts into the social relations and the material culture of families can be carried out. The contributors downplay past dichotomies between private and public spheres, placing a novel emphasis on the venues where transitions in domestic life dialogue with changing conditions at the level of the society. Some of the contributors to the volume discuss various acts of consumption through which the family shapes the home and its contents and thus negotiates meaning, identity, and status for itself and its members. However, such processes are also studied in reverse, with reference to the venues where the conditions of building and purchasing homes and their furnishings and the maintenance of existing property determine family life. One idea that emerges in this context concerns the contribution that the study of transformations in domestic spaces and their contents could make to our understanding of the changes within the extended family, between spouses and among siblings.

Our emphasis on the centrality of domestic consumption to the analysis of the dynamics of change affecting both the family and its relations with the wider society is also set in larger theoretical shifts. There is a move away from grand theories of modernization or world-system and dependency toward research that highlights process over structure, and/or cultural, economic, and political specificity over broad and over-uniform universal approaches.[6] This volume maintains that Middle Eastern families and their consumption patterns do not conform to simple notions of linearity and progress, according to which all kin relations, through time, are supposed to be restructured into those typified by the standard nuclear family engaged in an homogeneous modern lifestyle. Instead, the different studies of domestic consumption presented in this volume offer more complex accounts of various modernization experiences of families, in which these are observed constantly to readjust their most immediate and intimate surroundings. In parting from approaches that deal with uniform patterns of local consumption shaped by external dynamics, the volume further proposes a different reading of local material cultures, which highlights fluidity in the meanings of commodities in changing environments. This approach contrasts with world-system and dependency theories in both their highbrow and popular perspectives on globalization, such as McDonaldization, which stress a top-down and unidirectional impact of the West on the rest.[7] Instead it joins a growing number of studies on consumption in colonial and later developing countries (and recently in postcommunist countries) that argue in various ways that cross-cultural consumption is a complex phenomenon where resistance and agency of local consumers play a crucial role in determining the shape, scope, and meaning of observed consumption patterns.[8]

Many of the themes pursued by the contributors to the volume are also salient in recent literature on consumption in developed, Western countries. To a large extent, there is also a shared methodological sensitivity, which relates the literature in question to the chapters in this volume. These emphasize, for example, the significance of the home as a site for material culture studies that one finds in Daniel Miller's recent work.[9] Another clearly marked theme here, concerning the centrality of the home to the development and reproduction of social relations, occupies an important place in many contemporary currents of analysis of domestic consumption.[10] As opposed to the more static nature of structuralist analyses, which formed the core of earlier research, recent approaches assign much significance to "... the processes by which a home and its inhabitants transform each other."[11] This focus on process and motion in the study of the meanings of home material culture is also in line with a similar trend in Middle Eastern studies, and in studies on places outside the traditional West.

However, the contributions to this volume also reflect certain differences from other studies sharing similar methodological concerns in their attempt to go beyond the emphasis of symbolic interaction and the venues at which such interaction gives meaning to home and family life. Most contributors to the volume in fact quite explicitly attach significance to the political economic background to the processes that shape the meaning of the home and family life, even "behind closed doors." Put differently, the study of the material culture within the home proceeds without overlooking the impact of the state and the market on domestic consumption. It might be argued that this concern with political and economic factors simply results from the particular context in which homes are studied, namely developing countries of the Middle East as opposed to developed Western societies. The articles in our volume surely demonstrate that the heavy hand of political power and a centralized economy had much impact on the structure, content, and meaning of homes in the Middle East and on family life. However, a recent paper by Avner Offer suggests that even in democratic and free-market societies, the state has a major role in shaping the economy, and politics exerts much influence on consumption[12] (see also Ayşe Buğra, chapter 5, in this volume). This suggests that our emphasis on joining political economy to symbolic meanings of things in the study of material culture of the home should not be limited to places outside the developed world.

The Objectives and Outline of Chapters

Our volume foregrounds research into different venues in which politics (the state) and economics (the market) infringe on free choice and influence domestic consumption patterns. Chapters 1 and 2 by Lisa Pollard and Mona Russell respectively, show that the growth of the Egyptian state with its increasing intervention in day-to-day life enhanced the development of modern homes and new kinds of domesticity during the second part of the nineteenth and the early twentieth century. State intervention allowed two distinct paths. The first actively provided the necessary infrastructure for urban development and promoted public and private construction of new residential neighborhoods. Still more, starting with Muhammad Ali the ruling dynasty and other state elites gradually moved away from the extended *Mamluk*-style household into new homes and a less extended family structure. The second path, mainly achieved through education, promoted new visions of modern home and domestic life, because state officials considered the new home, with its novel architecture and hygienic practices, to be the template of modernity and therefore the fundamental building-block of the nation. State officials also promoted new forms of domestic life—a new division of labor between men and women, and children's education within the home—in order to "advance" the state. Moreover, the Egyptian national movement, while becoming a growing opposition to the colonial state, shared these visions of home and domesticity and identified them with the modernization of the nation as a central goal in the struggle for independence.

New forms of representation further informed the introduction of new homes, home durables, and notions of domesticity. Lisa Pollard discusses the venues where the novel state education system and the state printer with its sponsored books, newspapers, and magazines made the state a major promoter of new lifestyles in the home. This was accompanied by the development of commercial printing, which opened a new public arena and enabled Egyptian intellectuals from the national movement to join in this campaign. Russell examines a complementary development in which advertisers used the print media to sell commodities associated with such modern environments. Thus the state, the national movement, and the market came together, albeit for different purposes, in their representation of the new home and its content. As demonstrated by both Pollard and Russell, these new forms of representation further promoted novel definitions of personhood in Egypt.

New representations of modern homes and forms of domesticity were not, however, limited to the top-down manipulations of the state, the national movement, or advertisers. Indeed, Micklewright's and Laden's essays in this volume (chapters 3 and 4) both demonstrate reciprocity, agreement, and tension between contemporary political and economic ideologies and domestic life. Micklewright's chapter (chapter 3) is an interesting account of the ways in which the camera first entered Ottoman houses and enabled consumers to create their own images of family life and the home. The utilization of this imported medium imbued domestic life with novel local meaning and identity. Her account thus stands in contrast to the prevalent study of Orientalist European photography (and painting), in reinterpreting the camera and its products in a local rather than a global context. Laden's essay on domesticity (chapter 4) in the Israeli women's magazine *la-'Isha* further suggests a more open-ended dialogue between official and private representations of domestic consumption and family life. Her research centers on Israel in a period of forced state austerity and a strong ideological emphasis on the need to mold the country and the nation while sacrificing personal consumption and individual lifestyle preferences. Laden's nuanced account shows that while *la-'Isha* at times created an alternative consumerist vision of the Israeli home, domesticity, and a "bourgeois" lifestyle in and outside the home, it did not encroach on compliance with national politics and the existing economic and sociocultural system.

Ayşe Buğra's essay (chapter 5) studies the historical dynamics of social change in its cultural and political dimensions by focusing on the interplay between public policy and private consumption. Here, it is the lack of systematic government intervention that determines the nature of Turkish real estate and home appliances markets. By adopting an analytical framework that draws on Karl Polanyi's work, Buğra compares the significance of state redistributive processes that have shaped consumption patterns in Western developed countries in the post–Second World War era with the centrality of reciprocity relations in republican Turkey. She traces the role of reciprocity in Turkey by investigating the development of the housing market in general and irregular urban settlements in particular. She further demonstrates how similar mechanisms, built around reciprocity relations, have also influenced the functioning of sales networks developed by the large companies controlling the markets for household durables. Along the same lines, Tania Forte's essay on construction and consumption of Arab homes in the Galilee (chapter 6) illustrates how family-based access to credit, manpower, and building materials creates distinct domestic spheres and family life for Palestinians in contemporary Israel. Her account emphasizes the translation of power relations, both outside the home (between the Jewish state and its Palestinian citizens) and inside the family sphere (between the extended and the nuclear family and within the latter), into the shape and content of the house and their meaning.

Any discussion on changing consumption habits necessarily touches upon questions of nonconsumption or unequal consumption. Indeed, the home and its contents constitute the largest consumption expenditures for the majority of households. They therefore serve as a good index of existing gaps between different social groups within any given society. Tally Katz-Gerro investigates inequality in household consumption in Israel (chapter 7) by analyzing up-to-date statistics on consumption among different social groups. She quantifies this gap and exposes unequal access to commodities and qualifies it in meaningful ways, which contextualize household inequality within the unique consumption patterns of various ethnic, religious, and national groups in Israel. Her analysis displays for us the mechanisms through which consumption-based stratification occurs and the extent to which it echoes or crosscuts other social cleavages. It further calls for future research into the domestication of such social cleavages in various material cultures of families in Israel.

From the foregoing it should be clear that through our case studies we hope to provide some guidelines rather than a model for future research into domestic consumption and its impact on family life. Moreover, the volume is bound to be incomplete—we still need much research on the development of domestic consumption in other parts of the Middle East and under the impact of varying conditions such as oil revenues (in both producing and nonproducing countries), religious resurgence, wars, and globalization.

Nevertheless, we hope that the volume will contribute, both theoretically and empirically, to a better understanding of the subtleties required from future study in emphasizing the synergy between economic and political action and the physical and symbolic shape of homes and domestic life. We likewise argue for a less sharp distinction between private and public spheres, and for historicity and emphasis on process rather than rigid structural analysis of domestic consumption and the material culture of homes. Furthermore, we suggest that future research focus on the dialectics between humans and things in the study of the ways in which the family shapes its most immediate environment and is in turn shaped by it.

Notes

1. Alan Duben and Cem Behar, *Istanbul Households: Marriage, Family, and Fertility, 1880–1940* (Cambridge: Cambridge University Press, 1991); Beshara Doumani, ed., *Family History in the Middle East: Household, Property, and Gender* (Albany: State University of New York Press, 2003).
2. Jane Hathaway, *The Politics of Households in Ottoman Egypt: The Rise of the Qazdağlis* (Cambridge: Cambridge University Press, 1997); Margaret L. Meriwether, *The Kin Who Count: Family and Society in Ottoman Aleppo, 1770–1840* (Austin: University of Texas Press, 1999); Leslie P. Peirce, *The Imperial Harem: Women and Sovereignty in the Ottoman Empire* (New York: Oxford University Press, 1993).
3. This footnote cannot adequately cover the burgeoning of gender studies in the Middle East. The following collections are some examples of the literature that influenced my understanding of this field: Lila Abu-Lughod, ed., *Remaking Women: Feminism and Modernity in the Middle East* (Princeton: Princeton University Press, 1998); Amira el-Azhari Sonbol, ed., *Women, the Family, and Divorce Laws in Islamic History* (Syracuse: Syracuse University Press, 1996); Suad Joseph and Susan Slymovics, eds., *Women and Power in the Middle East* (Philadelphia: University of Pennsylvania Press, 2001); Nikki Keddie and Beth Baron, eds., *Women in Middle Eastern History: Shifting Boundaries in Sex and Gender* (New Haven: Yale University Press, 1991).
4. Ali Karman Asdar, *Planning the Family in Egypt: New Bodies, New Selves* (Austin: University of Texas Press, 2002); Carla Makhlouf Obermeyer, *Family, Gender, and Population in the Middle East: Policies in Context* (Cairo: American University in Cairo Press, 1995); Ismail Sirageldin, ed., *Human Capital: Population Economics in the Middle East* (London: I.B. Tauris, 2002).
5. Of special importance here is Unni Wikan's groundbreaking work: "Living conditions amongst Cairo's poor," *Middle East Journal*, 35, 1 (1985): 7–26. See also her *Tomorrow, God Willing: Self-made Destinies in Cairo* (Chicago: University of Chicago Press, 1996). Another significant writer in this field is Diane Singerman,

Avenues of Participation: Family, Politics, and Networks in Urban Quarters of Cairo (Princeton: Princeton University Press, 1995); see also her collection (with Homa Hoodfar), *Development, Change, and Gender in Cairo: A View from the Household* (Bloomington: Indiana University Press, 1996).

6. The literature on both Modernization and Dependency theories is huge, and encompasses a variety of disciplines including anthropology, development studies, history, politics, and sociology, each emphasizing different aspects of the two paradigms. Two of the most quoted works inspired by Modernization theory are Daniel Lerner, *The Passing of Traditional Society,* modernizing the Middle East (New York: Free Press of Glencoe, 1964) and Bernard Lewis, *The Emergence of Modern Turkey* (London: Oxford University Press, 1965). In scholarship on the history of the Middle East, Modernization usually went together with the literature on the so-called decline of the Ottoman Empire: see its critique in Roger Owen, *The Middle East in the World Economy, 1800–1914* (London: I.B. Tauris, 1993), Introduction.

   World-System/Dependency theory was inspired by Imanuel Wallerstein and the work of Turkish scholars; see especially: Huri Islamoglu-Inan, ed., *The Ottoman Empire and the World-Economy* (Cambridge: Cambridge University Press, 1987); Reşat Kasaba, *The Ottoman Empire and the World Economy: The Nineteenth Century* (Albany: State University of New York Press, 1988); Şevket Pamuk, *The Ottoman Empire and European Capitalism, 1820–1913: Trade, Investment, and Production* (Cambridge: Cambridge University Press, 1987).

7. Such literature on cross-cultural consumption has its theoretical roots in neo-Marxist literature, especially of the Frankfurt school, and the work of French writers such as Baudrillard and Guy Debord. It was recently transplanted into empirical research on colonial and later developing countries in works such as Timothy Burke, *Lifebuoy Men, Lux Women: Commodification, Consumption, and Cleanliness in Modern Zimbabwe* (Durham: Duke University Press, 1996).

   The broad public reception of George Ritzer's *The McDonaldization of Society: An Investigation into the Changing Character of Contemporary Social Life* (Newbury Park, CL: Pine Forge Press, 1993) and Naomi Klein's *No Space, No Choice, No Jobs, No Logo: Taking Aim at the Brand Bullies* (New York: Picador USA, 1999) represent the recent spread of such ideas from academia and journalism into political activism against globalization and its perceived institutions.

8. Some examples for this literature are: Arnold J. Bauer, *Goods, Power, History: Latin America's Material Culture* (Cambridge: Cambridge University Press, 2001); Deborah S. Davis, *The Consumer Revolution in Urban China* (Berkeley: University of California Press, 2000); Maris Boyd Gillette, *Modernization and Consumption among Urban Chinese Muslims* (Stanford: Stanford University Press, 2000); David Howes, ed., *Cross-cultural Consumption: Global Markets, Local Realities* (London: Routledge, 1996); Benjamin Orlove, ed., *The Allure of the Foreign: Imported Goods in Postcolonial Latin America* (Ann Arbor: University of Michigan Press, 1997); Donald Quataert, ed., *Consumption Studies and the History of the Ottoman*

Empire, 1550–1922: An Introduction (Albany: State University of New York Press, 2000). On consumption in postcommunist countries see a special issue of *Ethnos* 67: 3 (2002). See also: Susan E. Reid and David Crowley, eds., *Style and Socialism: Modernity and Material Culture in Post-War Eastern Europe* (Oxford: Berg, 2000).

9. Daniel Miller, ed., *Home Possessions: Material Culture and the Home* (Oxford: Berg, 2001). Although Miller refers to his volume as anthropology and the ethnography of studying the home, his broader theoretical analysis in the introduction covers other fields, especially history, which makes it good reading on the state of the art in research on homes.

10. Of the recent literature on homes see also: Leora Auslander, *Taste and Power: Furnishing Modern France* (Berkeley: University of California Press, 1996); Donna Birdwell-Pheasant and Denise Lawrence-Zuniga, eds., *House Life: Space, Place and Family in Europe* (Oxford: Berg, 1999); Inga Bryden and Janet Floyd, eds., *Domestic Space: Reading the Interior in Nineteenth Century Britain and America* (Manchester: Manchester University Press, 1999); Janet Carsten and Stephen Hugh-Jones, eds., *About the House: Lévi-Strauss and Beyond* (Cambridge: Cambridge University Press, 1995); Tony Chapman and Jenny Hockey, eds., *Ideal Homes? Social Change and Domestic Life* (New York: Routledge, 1999); Sharon Marcus, *Apartment Stories: City and Home in Nineteenth-Century Paris and London* (Berkeley: University of California Press, 1999).

11. Miller, *Home Possessions*, 2.

12. Avner Offer, "Why Has the Public Sector Grown so Large in Market Societies? The Political Economy of Prudence in the UK, c.1870–2000," *Discussion Papers in Economic and Social History*, University of Oxford, 44, March 2002.

PART I

The State, the National Movement, and the Modern Home

CHAPTER 1

Working by the Book: Constructing New Homes and the Emergence of the Modern Egyptian State under Muhammad Ali

Lisa Pollard

In the early nineteenth century, as Egyptian viceroy Muhammad Ali (r. 1805–1848) undertook the project of transforming the Egyptian state from confederation-based politics to a nation-state governed by a royal family, a set of images began to circulate amongst upper-class, learned Egyptians about the relationship between the domestic behavior of the Egyptian ruling elite and the ability of the nation-state to function. Texts connecting various sorts of domestic habits and customs with corresponding political systems worldwide illustrated that the rise of successful nation-states accompanied the transformation of a people's domestic habits from backward and barbarian to modern and civilized. In such texts, monogamy and ordered domestic relationships were the hallmarks of modernity and civilization. Texts produced by Muhammad Ali's state, through his school of translation (Dar al-Alsan) and his project of sending student missions abroad, created a map of the modern world on which each nation could locate itself according to its domestic and marital habits. State-produced knowledge about the world thus helped establish new conceptual relationships among the Egyptian elite between geography, ethnicity, culture and history in an Egyptian national consciousness that was becoming both increasingly global and markedly fixed at the local level.[1]

While such texts were not alone responsible for the transformation of Egyptian upper-class domestic habits over the course of the nineteenth century, and do not constitute the sole vehicle through which Egyptian elites began to discuss the question of modernity and what it would mean for Egyptian society, state-produced literature did play a role in shaping the political sensibilities of a generation of Egyptian state functionaries. While Muhammad Ali did not produce the modern household or link it to national identity, his state-building projects did serve as a kind of industry through which the modern, European household—with all its trappings—was introduced into the Egyptian market.

To reform the state, Muhammad Ali is claimed to have defined what modernity meant and how he wanted it implemented into Egypt.[2] At the same time, he appears to have isolated institutions and ideologies that did not meet his criteria for modernity and subjected them to reform. In both cases, the very act of refashioning Egypt depended on the state's active production of knowledge about the sciences and structures that it wished to construct and emulate, or dismantle and reject.

To know and construct the ideologies and institutions that were identified as having produced the European polity, Muhammad Ali created state agencies to penetrate and chronicle European institutions in ways that were often as thorough as Western travelogues about Egypt produced at the same time.[3] As the result of such projects, state servants became unwitting voyagers to hitherto unexplored terrains. Many of them thus "journeyed" without leaving Egypt. Others actually ventured abroad in search of new fields of knowledge. While that production of knowledge was not necessarily designed to dominate Europe, it did expose the arenas in which Europe had excelled.

Typically, inquiries into travel literature and translations from this period have concentrated on how contact with Europe changed the literary styles of Egyptian authors.[4] My interest here is not at all with style; rather it is, on the one hand, with the content of state-produced texts and, on the other, with the circulation of state-produced literature and its influence on reading audiences. The positivist view of history that such texts contained served as a new intellectual commodity in its own right. At the same time, however, state-produced ethnologies, in fixing the household as a measure of modernity and national identity, made the modern domicile, its contents, and the behavior of its inhabitants into intellectual and material commodities as well.[5]

What was distinct about state-produced, nineteenth-century Egypt "travel literature," was both its intended utility as part of the state's mission to transform itself, and the ways in which it exposed state servants to new ways of thinking about their personal behavior and Egypt's place in the world. Government institutions produced literature that put the world on tour, ranked the world's "nations" scientifically, and placed Egypt in a hierarchy of development at the apex of which sat "modernity." In such cartographies, the intimate details of domestic activities stood out as prominent features, and were used as units of measurement.[6]

The home, the family, and domestic behavior fully appeared as metaphors for Egyptian nationalism, political commentary, and critique in the years after the British occupation of 1882. Political and economic transformations throughout the nineteenth century were reflected in changes in ideas about the domestic realm and familial practices. The British brought with them to Egypt a discourse in which Egyptian marital and domestic practices signaled despotism and political backwardness, but Egyptians had already begun to view changes in household arrangements as heralding political and economic centralization and modernization. Reformed institutions, such as the household, became crucial markers on the landscape of modern Egypt.[7]

The House of Muhammad Ali and the Making of a Modern State

Muhammad Ali ruled Egypt through innovation and imitation. From the Mamluks who preceded him he adopted the practice of state centralization. He departed from Mamluk rule, however, in his agricultural and industrial projects, the building of a modern military and the creation of a nascent system of state-sponsored, secular education.[8] Muhammad Ali also gave birth to a cadre of civil servants who, over the course of the nineteenth century, became increasingly loyal to the Egyptian state and vested in its success. "The viceroy ruled absolutely, but with the help of nobles and technocrats he called into being."[9] In contrast to the structure of the Mamluk dynasties that had previously ruled Egypt, as of 1805 political power was embodied in one man and one household.[10] Unlike the houses of the Mamluks who preceded him, Muhammad Ali's household was able to command hegemony over Egypt by 1811.[11] To further consolidate his power, Muhammad Ali took land from Mamluk families and from the local religious elite—the `ulama— and placed it under state control. Among other things, land acquisition allowed Muhammad Ali to create a loyal elite. The viceroy gave tracts of land to members of his family, who while never learning Arabic became at least in part Egyptian as the result of their acquisition of Egyptian real estate. He also meted out land to the Ottoman elite who had developed ties to Egypt through service to the Ottoman state. To break their old attachments, former Ottoman officials were made landlords by Muhammad Ali. Land also served as a means of cultivating the loyalty of Egyptian notables, who through the acquisition of large parcels of land became invested in Muhammad Ali and his projects.

Muhammad Ali also used his fledgling state-run educational system to cultivate loyalty in the sons of this new, landed notability. In its early days, Muhammad Ali's state resembled a Mamluk household, as the viceroy placed a substantial amount of power in the hands of his family members. But as the state grew and expanded, so too did its need for functionaries. Hence, a second avenue into Muhammad Ali's "family" of military and administrative officials became the nascent educational system.

The roots of the Egyptian state educational system lie in the vice-regent's desire to reform the military and to make it fully loyal to him. Muhammad Ali replaced the Mamluk system through the conscription of ordinary Egyptians into his military, and through the establishment of an European-trained officers' corps. Immediately after securing his power, the viceroy began relying on Europeans to train his new military elite. Egyptians' lack of acquaintance with European languages made it necessary for Muhammad Ali to send Egyptians to Europe and to bring Europeans to Egypt to teach foreign languages. Language instruction led to the establishment of a small number of military schools for the sons of the notable class (both Arabophone Egyptian and Ottoman-Turkish Egyptian) in which a Western, secular education replaced traditional Quranic learning. Graduates of these schools took jobs in the military or became members of the fledgling corps of civil servants who provided the foundation for an expanding administration. Between 1809 and 1849, 11,000 Egyptians passed through Muhammad Ali's schools.[12]

The creation of a cadre of state-trained civil servants contributed to the beginnings of an Egyptian identity amongst those who worked for the state.[13] The acquisition of land by Muhammad Ali and his descendants and by the notables and state servants to whom land was granted over the course of the nineteenth century did much to cement the relationship between the ruling dynasty and the territory known as Egypt. But the creation of an administrative culture, shaped by the state and circulated by the movement of landowning notables and their sons between Cairo and the provinces led to the rise of a rank of Egyptians with increasingly common interests. Additionally, a new, professional class of state-trained bureaucrats who did not necessarily own land—the *effendiya*—came to share in this new culture. As the century progressed, the culture that was shaped by land and vested interests in running the state began to lessen the distinction between Arabophone Egyptians and their Ottoman-Turkish-speaking counterparts.[14]

Bringing the Nation and the Household Together

Muhammad Ali is credited with having said that translation, and then the printing of translated books, were the best means of carrying out his goal of exposing Egypt to Western sciences and culture.[15] From 1835 to 1849,[16] his school of translation (known initially as Madrasat al-Tarjama [The School of Translation] and then as Dar al-Alsan [literally, The House of Tongues]) produced translations on topics ranging from science and medicine to history, geography, philosophy, and logic. The original purpose of the school was the production of better bureaucrats for the state administration. Its goals later expanded to take on the creation of a core of translators who could make knowledge accessible and useful to future generations.[17] Ultimately, Dar al-Alsan's main business was the dissemination of Western ideologies, institutions, and innovations.[18] By the time of Muhammad Ali's death, approximately 1,000 texts had been translated from European languages into Turkish and Arabic.[19]

Egyptian historian Ahmad `Izzat `Abd al-Karim thus refers to the era of Muhammad Ali's rule as that of "translation and Arabization" (*asr al-tarjama wa-l-ta`rib*). Al-Karim claims that Muhammad Ali believed that "modernity" and all its useful devices had been written about by those who invented them; through his translation project, he would apply those devices to his own country.[20] Translations produced by Dar al-Alsan over the decades of its existence and published by state-controlled presses in Bulaq[21] and Alexandria formed the staple diet of Egypt's literate classes, and were instrumental in the formation of a new Egyptian intellectual elite.[22]

In the 1820s, before translation was officially taught and while the Egyptian state was still relying on expatriate Syrians as well as the student missions abroad to provide it with translators,[23] Muhammad Ali was interested in histories about men (and, sometimes, women) who had reformed their nations. The viceroy is said to have told the story of a certain Frenchman, Colonel Duhamel, who informed him that he would only become a great man if he read history.[24] Accordingly, he had biographies of the prophet Mohammad, and rulers such as Alexander the Great, Catherine the Great, and Napoleon translated into Turkish. Some of these texts were later published and circulated: In 1824, for example, Machiavelli's *The Prince* (translated from Italian into Arabic as *al-Amir fi `ilm al-tarikh wa-l-siyassa wa-l-tadbir* or, *The Prince (as he is known through) the Science of History, Politics and Organization* was published by Dar al-Kuttub al-Misriyya. The following year, Castera's *Histoire de l'Imperatrice Cathrine II de Russie* was published by the Imprimerie de Bulaq.[25]

In the years after the founding of Dar al-Alsan, graduates continued to make European history available in Turkish and Arabic. In the 1840s, Rifa`a al-Tahtawi (1801–1873), administrator and translator at Dar al-Alsan from 1835–1849, published *Tarikh dawlat italia*, his translation of "A History of the Italian State." In the 1840s, the Bulaq publishing house printed translations of a number of histories written by Voltaire about European rulers such as his *Lion of the North: A History of Charles XII of Sweden*,[26] and *Life of Peter the Great*, which was translated into Arabic in 1842. Such texts depicted history as being driven by the wills of great men, and the nation-state as the product of their reform programs.

The translation and production of history by civil servants for other civil servants and for the consumption of a reading public served to place Muhammad Ali and his state-building projects in the company of other rulers whose greatness, personality, and character were evidenced by the reform projects that they gave birth to.[27] Translations also attached success-ful reform programs to the private habits of great men and women, as is illus-trated in an 1841 translation, *Nazam al-lal'i fi al-suluk fi man hakam fransa min al-muluk* (roughly translatable as *The Superlative Behavior of the Kings who have Governed France*), which equates the ability to create a strong state with particular kinds of habits and behavior. The history of the formation of the modern nation-state was thus charted along the variables of habits and customs, especially those of rulers.[28]

This ethnography *qua* history is well illustrated by al-Tahtawi's 1833 trans-lation of Frenchman Georges-Bernard Depping's *Aperçu historique sur les moeurs et coutumes des nations*,[29] which was a staple text in the Egyptian pub-lic school system for the next century.[30] Al-Tahtawi's translation of Depping was divided in two parts. The first, called "On the Needs of Humans, and the Way they Live with their Families and their Clans," catalogued nations in terms of their houses and the customs practiced in them; their clothes; clean-liness; marriage and its various customs; women; offspring; old age; funerals; hunting, both on land and sea; commerce and money.[31]

Part two, "On Morals and Customs in Relation to Nations and their Mentalities," considered the effects of morals and habits on the shape of nations and on their progress. It contains chapters on topics like games and sports, poetry and music, writing, literature and laws, dance, hospitality with guests, holidays and seasons, slavery, beliefs, and politics.

In a chapter from part one on "cleanliness," Depping compared the civilized to the uncivilized peoples saying:

In non-Muslim lands[32] there is a great preoccupation with the cleanliness of objects, especially with that of the domicile. The cleanest are the

Dutch: in their cities you see that most streets are paved with cobblestones, ordered and clean. Their homes are beautiful on the outside, and their windows are washed.... One finds a certain amount of cleanliness among the English and some of the Americans. The French aren't very clean, nor are the Germans.... The uncivilized peoples are filthy. Among them you find many who have lice, and many who eat their lice. They smear their bodies with lard. Lots of people in the Americas and Asia spend a good deal of the year underground, where the air is plenty unwholesome. The foul air combines with other putrid odors, like those of their food. These people live for long periods without cleaning anything... they eat meat with their hands, and never touch a fork, knife or spoon.[33]

Here, a hierarchy of cleanliness corresponded to a hierarchy of political organization and development. To Depping, modern nations were the nations that kept house. About his own country, he wrote: "If the French are given to the sciences... if they enjoy a prosperous economic life and continue to cultivate refined habits and morals, this is thanks to favorable political order."[34]

In Asia, one of the "filthy" regions, such political leadership had not really developed. There were often strong kings, but no order, especially not the kind of order that would lead to controlled agricultural and industrial production as it was found in parts of Europe. He said:

In Asia, one finds both strong kings (*muluk*) and sultans (*salatin*). Most of them rule however they please; they spill the blood and spend the money of those over whom they rule in any way they like. Power is had through the gaining of favors from the ruler.

Most of the rulers only rarely leave their *harems*, so it is hard to get an audience with them. They never hear the petitions of those who have been poorly treated.... One of the habits of the citizens of Asia is to hide money from their rulers; it is the habit of the rulers, in their public addresses and their *firmans*, to compare themselves to the sun and the moon. Such rulers never go out unaccompanied by their military, and their houses resemble small cities.[35]

Political and historical development are thus underscored by investigation into the living conditions of both kings and subjects. While Depping gives no indication of how nations transcended putrid-smelling living conditions to reach neat, orderly streets, his implicit message is that it was only in doing so that centralization and industrial development would take place.

In the 1830s, translations of texts on geography—which was defined as being a science through which nations could be ranked and ordered[36]— reinforced the notion that the globe was divided into civilized and uncivilized, clean and unclean nations. During the 1830s, a number of French texts on descriptive geography; cosmography; physical geography; religious, political and historical geographies, as well as moral geographies were all translated into Arabic, including al-Tahtawi's translation of Malte-Brun's *Géographie universelle* in 1838. Translations of geography were influential in producing a new corps of Egyptian geographers, and in creating interest in geography as a new practice.[37] A review of primary- and secondary-school literature from 1901 shows that al-Tahtawi's translation of Malte-Brun (*al-Jeographiyya al-`umummiyya* in Arabic), continued to be used in Egyptian classrooms for many years.[38]

Malte-Brun's geography was both literal and theoretical.[39] His theoretical sections present geography as a literal system through which the world is divided, measured, and ordered; through which the reader "sees" the contents of the globe, and locates the multitude of nations by which he is surrounded.[40] "Is not geography the sister and the rival of history? If the one enjoys the empire of universal time, does not the other rightfully claim that of place?"[41] Malte-Brun's "precise science" also extended to empirical discussions of the habits and customs—both public and private—of nations; knowing the intimate habits of the globe's inhabitants was added to the criteria through which the "universal system" was applied. His general "theory of geography" insisted that the world could be known through a study of morals, political tendencies, and national "character," and hence the last "book" in his theoretical overview was dedicated to the study "Of man considered as a Moral and Political Being; or Principles of Political Geography."

Malte-Brun's tendency was to divide nations not by the natural phenomena that separated them, but, rather by the languages they spoke, their religion, the forms of their government, the way they ate and drank, and the homes they inhabited. Malte-Brun's geography was intended to be a science that would cover the globe, from its outer crust to its very inner core, with the task of banishing the monsters of the world outside Europe's border, and converting "space into place."[42]

> . . . we shall take a view of the leading features of nature . . . . We shall seek our way downward . . . and thus do our utmost to explore the structure of the globe. We shall conclude the picture by considering man in his natural and in his political condition. We shall classify the races of our species according to the varieties which are marked in their bodily appearance

and character—according to the languages which they speak—according to the creeds by which their minds are consoled, or degraded and enslaved—and according to the laws which mark the progress of civilization, or the profound darkness of utter barbarism.[43]

Malte-Brun's new geography can therefore be characterized as the science through which nations were categorized according to the natural phenomena that separated them, and through which the barbaric peoples were distinguished from the civilized through an empirical, scientific analysis of manners and morals, customs and proclivities. Like its longitude and latitude, the behavior of a region's inhabitants, both ruler and ruled, its topography, and its flora and fauna, were crucial to a nation's ranking in Malte-Brun's ordered, universal system.

What Malte-Brun called an "exact" science was in fact a subjective cataloging of morals, habits, and customs—a long series of "ethnographies." This formula, according to the author, was a kind of concession, in which science and subjectivity would collapse on one another when such an implosion was crucial to the process of *knowing* the world, nation by nation.

We shall not even scrupulously deny ourselves and our readers the *pleasure* of occasionally mingling our topographical descriptions with . . . anecdotes to illustrate manners, often serving to fix in the memory names of localities, which otherwise it would be difficult to retain.[44]

This treatment of geography produced a cosmology of national boundaries that were both fixed and fluid. Nations would continue to be located by the natural phenomena by which they were boundaried and bordered. They were placed, however, according to the habits of their inhabitants—habits that were subject to change.

The Egyptian State Abroad

To begin building a state and a new educational system, Muhammad Ali imported advisers and instructors from Europe. It quickly became more practical and less expensive, however, for him to send students and young bureaucrats to Europe to learn the languages and the sciences that had previously been taught by Europeans in Cairo. It became "typical" of Muhammad Ali, ". . . to send his own kind to Europe to see for themselves what was lacking in the country and what the Westerners had to give and teach."[45] As early as 1809, he began sending students to Italy to learn how to establish munitions factories, build military arsenals, and other technical facilities. Their job would later be to teach other Egyptians to do the same.[46] Milan, Rome, and Florence hosted Egyptian students who studied military science, shipbuilding, engineering, and printing. A second group of approximately thirty Egyptian students was sent to England in 1818 to study shipbuilding and mechanics. There appears to have been no missions between 1818 and 1826, when the mission that included Rifa`a al-Tahtawi left for France.[47] Because of its size and its superior organization, that mission is often referred to as the first real mission abroad.[48] Upon their return to Egypt, the mission's members were given state positions in translation, civil administration, finance, education (*diwan al-madaris*) and in various forms of industry and established themselves as a kind of new, cultured aristocracy.[49]

Members of Muhammad Ali's student missions to Europe in the 1820s and 1830s were given a copy of another of Depping's texts, *Evening Entertainment*, to read as part of their basic curriculum.[50] *Evening Entertainment* was a fictive *tour du monde* in which a father recounted his adventures as a world traveler to his children. The purpose of the book was to expose students to the "precise sciences" of geography and history as they embraced the study of the whole globe. The text included biographies of figures like Peter the Great, histories of reform programs, and catalogues of each nation's habits and customs—both public and private.

Given the templates with which they were educated, it is not surprising that students who were sent to Europe by the Egyptian state throughout the nineteenth century and who recorded their journeys were concerned with both the physical and moral local location of both Europe and Egypt. Like Dar al-Alsan, the missions had the goal of amassing and producing knowledge about the world outside of Egypt; students abroad studied not only Europe's languages but also the institutions through which Europe claimed its modernity.

The memoirs of the most celebrated example of them, al-Tahtawi,[51] became not only a celebrated piece of Arabic literature, but a seminal work on "modernity" and what it would require of Egyptians. *Takhlis al-ibriz fi talkhis baris* (usually translated as *An Extraction of Gold in a Summary of Paris*), which was published by order of Muhammad Ali in 1834,[52] was apparently written to tell the Egyptians as much as possible about European society and its customs, and to give suggestions for shaping Egyptian institutions along French lines. He wrote the book because: "I intend to exhort the Islamic countries to look into foreign sciences and research; it is well known that they have been perfected in Europe.... I wanted them to see the things of which the Islamic world had been deprived."[53] Muhammad Ali had free copies given to his state functionaries as well as students in the government schools.[54]

Literary critic Roger Allen places *Takhlis al-ibriz* in what he calls the Egyptian "grand tour" literature of the mid- and late nineteenth century.[55] Like state functionary 'Ali Mubarak's *'Alam al-din*, published later in the century, *Takhlis al-ibriz* was an entertaining and palatable account of a foreign country through which Egyptians could "see" how a "modern" society was structured and how it worked. Unlike Mubarak's work, which was a fictional account written as a series of conversations between a shaykh from al-Azhar and an English Orientalist while they toured Europe, al-Tahtawi presented his work as a "factual" account of what he observed during his travels in France. In both cases, however, tours through Europe and its institutions, both real and imaginary, were the vehicle through which French society was "exposed": both men used visits to European institutions and excursions through public offices, museums, public libraries, hospitals, banks, the Bourse, theaters and, finally, private homes to let Egyptian readers "see" Europe.

While *Takhlis al-ibriz* was a personal account of al-Tahtawi's impressions and experiences in France, the work in many senses conforms to the conventions of Depping and Malte-Brun's "science."[56] Like the geographers, al-Tahtawi began his summary of France with a discussion of how France could be *located*. Accordingly, the first section of the book's third section, "A Description of Paris and its Civilization," is called "On the Geographic Placement of Paris, and the Customs of its People." France's exact location on the globe, its longitude and latitude, its distance from other parts of the world including Egypt and its resulting weather; France's topographical and geographic features, such as its many mineral springs; its flora and fauna— all appear to be requisite to the reader's understanding of where France was located, and what that placement produced in terms of its "French-ness."

Al-Tahtawi's quest to "scientifically" locate France quickly gave way to lengthy discussions of the habits and customs of its inhabitants. France's social and political institutions were thus uncovered and analyzed. Most of the book's third section covers such topics as "The Habits of the Parisians in their Homes," "Food and Drink of the Parisians," "The Clothing that Parisians Wear." Al-Tahtawi also included a chapter on parks and places of diversion.

In some senses, the chapters in which al-Tahtawi focuses on French institutions and the habits of the French in them do not differ from nineteenth-century European travel literature written about Egypt. Parisian homes, for example, a general description of the order of Parisian streets, the kinds of houses the different classes lived in, and the types of materials with which they were constructed all served to help Egyptians see and know the French.[57] But there was a second agenda, most akin to European literature about Egypt. Al-Tahtawi claimed that overviews of a city helped a reader surmise whether a people was civilized or barbaric.[58]

Al-Tahtawi then switched from the streets of Paris to the home of an upperclass Parisian, where he positioned himself as a surveyor of the home's contents, and as an observer of its activities. Like contemporary European travel literature al-Tahtawi's accounts record the home and its contents in extraordinary detail:

> In each room there is fireplace, made of marble, on the mantle (ledge) of which a clock is placed. On each side of the clock, one finds a vase made of marble or of faux marble, in which flowers (both real and fake) are placed. On either side of the vases sit candelabras . . . . In most rooms there is a musical instrument called a piano . . . . Most rooms have pictures in them.[59]

Al-Tahtawi draws the reader's attention to the fact that bourgeois Parisian home decor was a means of displaying a family's history: "Most rooms have pictures in them, especially of [the owners'] parents. Often an office will contain marvelous displays of art and of interesting objects which might well have belonged to [the owner's] ancestors."[60]

Likewise, al-Tahtawi used the contents of the rooms that he surveyed to draw conclusions about the interests of the Parisians and their proclivities. He wrote, for example:

> If the room is a workroom or a reading room there is a table on which writing implements are kept, such as paper . . . . In one room I believe I saw a table with different kinds of documents on it, and I think I also saw in the rooms of upper-class homes brilliant chandeliers, of the kind that are lighted up with candles. And I think I saw in rooms where guests are received, a table on which a pile of books and papers had been placed for the guests' perusal . . . . All these things lead me to believe that the French place great importance on reading.[61]

He remarked that French homes, in general, could be characterized by a remarkable degree of cleanliness and order (he compared them only to the peoples of the Low Countries, for example, saying that the latter were the cleanest and most orderly people in the world), and mentioned that Parisian homes got lots of light and air, contributing to the general well-being of their inhabitants.[62]

Of the domestic habits of modern Parisians, al-Tahtawi seemed to be most compelled by table manners. While he was initially somewhat startled by the presence of tables and chairs in Parisian dining quarters, it was the order of the dining room table itself that seemed to catch his greatest attention. He said:

> On tables are always placed a knife, fork and spoon of silver. For them [the French] it is a matter of cleanliness that one not touch the food with his hands. Each person thus has a plate in front of him, as well as a glass for drinking. . . . No one drinks out of another's glass. There are individual containers for salt and pepper. . . . Everything on the table is well ordered.[63]

Schools—from primary schools through "les academies"—were noted by al-Tahtawi as being "common" to the daily life of middle- and upper-class Frenchmen and to a certain percentage of French women. He made clear connections between schools, modern domestic habits, and the French interest in modern sciences. "Conservatoires," where the "tools" of astronomy and engineering were housed, "academies," "institutes," and "lycées" all formed al-Tahtawi's tour of the institutions through which the sciences were produced and propagated. He wrote: "The sciences progress everyday in Paris . . . not a single day passes without the French making a new discovery. Some years, they even discover a number of new disciplines, a number of new industries, a number of new processes."[64]

Al-Tahtawi noted the overwhelming availability of schools in which the French could pursue an education as well as the financial commitment of the French government to their education. He found the number of newspapers and journals—"papers printed each day"—through which the French were educated "outside of the classroom" to be praiseworthy, as he did the many varieties of journals and the number of people who read them.

When he finished his tour of Paris, al-Tahtawi paused to remind the reader of what they had "seen" as the result of his investigations into French institutions. What he had just visited, he said, in homes, schools and institutes, were the "sciences" of Paris, the basis of France's progress, the sources of its success. From those sciences, he claimed, had sprung a number of ideologies and practices:

> And from the sum of the sciences of Paris can be had almanacs; new censuses, and corrected [lists of] marriages[65] as well as things of that nature. Every year, there appear a number of almanacs which record new discoveries in the arts and sciences, and which record governmental matters, and

which list France's elite by name, address and profession. If anyone needs the name of one of these families, or to find their house, he just looks at the almanac.

In Paris, people have special rooms in their homes for reading, and there are public reading rooms where people go to learn, where they read all the latest newspapers, journals and books. They borrow the books they need, read them and return them.[66]

Science had produced a number of customs: annual discoveries, and lists made of them; new records of births and deaths, and government-produced lists of who married whom. From inside their homes, in which there were special rooms for the project of reading, the French could learn what had recently been discovered, and where married couples lived. State and citizen alike had access to the domestic affairs of other citizens and interested readers made inquiries into such affairs from inside their homes. The state—well organized and scientific—extended its hand into private and domestic affairs of its citizens, and made knowledge about those affairs public.

The Landscape of an Increasingly Modern Cairo

As state modernization projects continued, parts of Cairo's landscape began to resemble descriptions from al-Tahtawi's texts. While the most significant transformations of Cairo's terrain did not take place until the 1860s under Ismail's reign, the rule of Muhammad Ali and his successors witnessed the appearance of the markers of modernity that al-Tahtawi describes.

In 1808, Muhammad Ali began the construction of a European-style palace and gardens for himself in Shubra, just north of Cairo's ancient walls. He had a wide avenue constructed connecting the palace with the city. To further mark his household's rise to power, the Pasha turned his attention to the Citadel in 1812, ridding it of almost all the Mamluk buildings, replacing them with his own palace, army barracks, government buildings and, later, a mosque. The vice-gerent's modernizing agendas were also reflected in the appearance of cotton, silk, and paper factories in Cairo, its surrounding villages and the provinces, part of a boom in industry that lasted through the late 1820s.[67] Bulaq, just north of Cairo, saw the construction of textile factories as of 1818, a press in 1822, and a foundry in 1829. Schools were also opened in Cairo and Bulaq: the school of civil engineering opened in Bulaq in 1821, becoming a polytechnic institute in 1834. The Abu Za`bal medical school and hospital were opened in the north of Cairo in 1827. Dar al-Alsan opened in 1835.

In his later days in power Muhammad Ali concerned himself with Cairo's public works. He issued laws to clean, clear, and widen the streets. One European reported that in 1832, Cairo's streets were cleaned three times a day, and that garbage was collected and burned outside the city.[68] The viceroy banned the use of *mashrabiyya* (lattice) window coverings, which tended to hang out over the streets, and ordered them replaced with glass windowpanes. The reasons behind the prohibition of *mashrabiyya* are said to have been fire. Some historians have suggested, however, that the Pasha intended new building styles to mark a new era.[69] Cairo's transformation was further marked in 1845 when Muhammad Ali formed a council responsible for further ordering and organizing Cairo (*majlis tanzim misr*), giving it the task of naming the streets and numbering the buildings on them.[70] The council proposed opening up the city through the demolition of certain old neighborhoods and the subsequent construction of two wide boulevards—a project that was only completed under Ismail.

Muhammad Ali and his elites—both Ottoman and Arabophone Egyptian—began to reside in new dwelling styles. The vice-regent, his family members and, soon, the elite he brought to power built houses and palaces all over Cairo.[71] But they did not build in the style of their predecessors. When Muhammad Ali eliminated his Mamluks by killing the last of them off in 1811, he also eliminated their style of architecture. Mamluk houses, in which 150–200 people often resided, were not torn down; in fact, the viceroy presented former Mamluk estates to his family members as gifts. But new residential styles soon overtook the old, and housed one family rather than confederations of families. By 1850 an allegedly Greek home called the Constantinopolitan had become popular with Muhammad Ali's family and the new elite classes. According to `Ali Mubarak, the new elites started building the Constantinopolitan in order to imitate the royals.[72] (Mubarak himself lived in such a house.) By the second half of the nineteenth century, the popularity of this fashion was eclipsed by neoclassical and Rococo. All became symbols of the position and power of the new elite.

By the 1850s, the rise of Arabophone Egyptians to bureaucratic prominence, their acquisition of land, and their increasing desire to set themselves apart from civil servants with Turkish backgrounds gave rise to a new, Egyptian bureaucratic culture. "They [Arabophone Egyptians] felt obliged to live in a certain way, to buy the 'right kind' of home and furnish it appropriately."[73] Historian F. Robert Hunter argues that by the 1860s landowner-ship and living practices among the bureaucratic elite served to diminish the Ottoman-Turkish-Arabophone Egyptian split, replacing it with a common set of interests and a common "Egyptian" bureaucratic culture. While animosities still existed between the two groups over who had the right to rule Egypt, and while Arabophone Egyptian resented their practical exclusion from the highest echelons of military and civilian administrations, personal wealth gave rise to greater commonalties between them.[74]

Changes in Cairo's appearance were obviously not the simple reaction to texts produced by Muhammad Ali's education and translation projects; change did not just happen "by the book." Mandates for construction and development met with changes in the tastes and demands of a growing, increasingly prosperous Egyptian elite, who tended not to return their investments to their land but rather spent their profit on consumption and on a European style of living.[75]

Conclusion

In the decades following the opening of Dar al-Alsan and the institutionalization of the student missions abroad, this relationship between science, state, and citizen became a common ingredient in the shaping of upper-class Egyptian nationalism. The process of building the state in early nineteenth-century Egypt resulted in a particularly configured relationship between the state, its citizens, and the world outside of Egypt's borders. As an ethnographer—unwitting or otherwise—Muhammad Ali had produced a body of literature and blueprints about modernity and its trappings. In that literature, produced and published for the purpose of expanding the state and its dominion, the domestic realm of the citizen both at home and abroad was penetrated, chronicled, and used as a yardstick for measuring modernity. Through "travel literature," the state began to see itself as modern, and to envision for itself a position in the modern world. To write such a book or translate one meant buying oneself a place in the state's employ. Purchasing and reading state-produced literature meant exposing oneself to habits and customs that would increasingly be defined as Egyptian. Buying or constructing a modern home in a neighborhood designed for those in the state's employ, filling it with European furniture, and engaging in increasingly Western behavior—all signaled proximity to the state and membership in the culture it had created.

This culture, and the men who practiced it, were not isolated to the halls of the royal palace. Rather, their presence, influence, and activities extended outward into other arenas of public culture. Ahmed Hussein al-Rashidi, for example, who was the translator of *Géographie universelle*, taught and translated in the School of Medicine in Cairo. Likewise, Ahmed `Abid al-Tahtawi, who translated Voltaire's history of Peter the Great into Arabic, later became representative to Cairo's Council of Merchants (*wakil majlis al-tujjar*). Abdullah abu al-Sa`ud, responsible for the translation of *The Superlative Behavior of the Men who have Governed France*, was a prominent figure in the Bureau of Translations, taught at Dar al-`Ulum, a teachers' college founded later in the century, and, later, edited Ismail Pasha's palace newspaper, *Wadi al-nil.*[76] From Dar al-`Ulum, lessons on modern "mapping" and geography made their way into primary- and secondary-school curricula such that Egyptian schoolchildren became familiar with the relationship between domestic behavior and the location of their country.

Ultimately, the modernizing project created a new geography in which Egyptians came to know and view themselves. Confounding what Edward Said has referred to as the "imaginative geography" of Orientalism, in which the world was divided into East and West and assigned an according set of political and cultural characteristics,[77] the geography that was produced by the state and its servants seems to blur such distinctions and make "modern" an overarching category. That transnational geography of modernity insinuated Egypt into a hierarchy of nations that challenges categories such as "East" and "Oriental."

At the same time, this new geography was remarkably local. It created for the Egyptians who were exposed to it, a kind of social geography that evoked and required an increasingly common set of traits and behaviors. The territory known as modern Egypt began to be infused with new cultural codes, and the emergent nation-state began to be ordered through the politics of modern behavior.

Muhammad Ali most certainly did not have such uses for domestic behavior in mind when he unleashed "modernity the way he wanted it produced." But in the century following the inauguration of the student missions and the opening of Dar al-Alsan, texts on modernity had ceased to be merely a product created by the state for its own uses. The link between domestic behavior and the progress of nations became intellectual commodities that could be had in the schoolroom, the library, and the modern press. At the same time building materials for modern domiciles, fashions, gadgets for the home, and behavior became the commodities through which modern, bourgeois Egyptian nationalism was displayed. Forces unleashed by the state quickly escaped the state's control.[78] While the Egyptian state did not invent the modern household or politicize it, its role in producing a body of literature in which connections between politics and domestic behavior were linked should not be underestimated.

Notes

1. Historian Xiobing Tang calls this the "global space of modernity." See his *Global Space and the Nationalist Discourse: The Historical Thinking of Liang Qichao* (Stanford: Stanford University Press, 1996). See also Rebecca E. Karl, "Creating Asia: China in the World at the Beginning of the Twentieth Century," *American Historical Review* 103, 4 (1998): 1096–1118.

2. Jamal al-Din al-Shayyal, *Tarikh al-tarjama wa-l-haraka al-thaqafiyya* (Cairo: Dar al-Fikr al-Arabi, 1951), 45.

3. See, *inter alia*, Rana Kabbani, *Europe's Myths of Orient* (London: Quartet Books, 1986); Judy Mabro, *Veiled Half-Truths: Western Travelers' Perceptions of Middle Eastern Women* (London: I.B. Tauris, 1991).

4. See, *inter alia*, Roger Allen, *A Study of Hadith `Isa Ibn Hisham: Muhammad al-Muwaylihi's View of Egyptian Society during the British Occupation* (Albany: State University of New York, 1974) and his *The Arabic Novel: An Historical and Critical Introduction* (Syracuse: Syracuse University Press, 1982).

5. See, *inter alia*, Arjun Appadurai's introduction to his edited volume, *The Social Life of Things: Commodities in Cultural Perspective* (Cambridge: Cambridge University Press, 1986); Jean Baudrillard, *The System of Objects*, James Benedict, trans. (London: Virgo Press, 1996).

6. Here my thinking has been greatly influenced by Benedict Anderson's *Imagined Communities* (London: Verso, 1991), and Anne Godlewska and Neil Smith's *Geography and Empire* (Oxford: Blackwell Publishers, 1994). The most persuasive account of the project of "mapping" and knowing the world, as a tool of European dominance and national resistance is found, however, in Thongchai Winichakul's *Siam Mapped: A History of the Geo-Body of a Nation* (Honolulu: University of Hawaii Press, 1992). Winichakul argues that the Siamese nascent "state's" control over both political power and technology, in the mid-nineteenth century, led to the construction of Siamese "territoriality." Territoriality included the state's attempt to classify an area, to communicate that classification to its citizens through the construction and strict enforcement of boundaries. Intimate knowledge of the inner, domestic realms, he notes, appear on late nineteenth-century Siamese maps, just as locally produced maps also placed Siam in an international context. Locality, or territoriality, had both an international and a very personal dimension.

7. See my "The Family Politics of Colonizing and Liberating Egypt, 1882–1919," *Social Politics* 7:1 (2000): 47–79.

8. Afaf Lutfi al-Sayyid Marsot, *Egypt in the Reign of Muhammad Ali* (Cambridge: Cambridge University Press, 1984).

9. Juan R.I. Cole, *Colonialism and Revolution in the Middle East: Social and Cultural Origins of Egypt's `Urabi Movement* (Princeton: Princeton University Press, 1993), 28.

10. F. Robert Hunter, *Egypt Under the Khedives, 1805–1879. From Household Government to Modern Bureaucracy* (Pittsburgh: The University of Pittsburgh Press, 1984), 22.

11. Ehud Toledano, "Social and Economic Change in 'The Long Nineteenth Century,'" in M.W. Daly, ed., *The Cambridge History of Egypt* (Cambridge: Cambridge University Press, 1998), 256.

12. Hunter, *Egypt Under the Khedives*, 113.

13. This is Hunter's argument in *Egypt Under the Khedives*.

14. See Hunter, *Egypt Under the Khedives*, and Cole, *Colonialism and Revolution in the Middle East*. Ehud Toledano makes a compelling counterargument in his *State and Society in Mid-Nineteenth-Century Egypt* (Cambridge: Cambridge University Press, 1990) in which he argues that the distinction between the two groups remained rigid through the later decades of the nineteenth century. See especially, 68–93.

15. Jamal al-Din al-Shayyal, *Tarikh al-tarjama*, 200.

16. Muhammad Ali's successor, Abbas (r. 1848–1854), had the school closed. It was not opened again until the reign of Ismail (1863–1879).

17. Al-Shayyal claims that copies of Dar al-Alsan's translations were, initially, produced for students in the nascent, national educational system and in the colleges of medicine and engineering. One thousand copies of each text were printed, and students were instructed to hand down their books to students who came after them. Ibid., 224.

18. On the role of the state and its servants in the creation of a body of literature called "modern history" in nineteenth-century Egypt, see Jack A. Crabbs, Jr., *The Writing of History in Nineteenth-Century Egypt: A Study in National Transformation* (Cairo: The American University of Cairo Press, 1984).

19. Al-Shayyal, *Tarikh al-tarjama*, 147.

20. Ahmad `Izzat `Abd al-Karim, *Tarikh al-ta`lim fi `asr muhammad `ali* (Cairo: Maktabat al-Nahda al-Misriyya, 1939), 329.

21. Bulaq Press was founded by muhammad `ali in 1821, and printed its first book in 1822.

22. Louis `Awad, *Tarikh al-fikr al-misri al-hadith, min al-hamla al-faransiyya ila `asr isma`il* (Cairo: Maktabat Madbouli, 1987).

23. See al-Shayyal, *Tarikh al-tarjama*, chapter two.

24. Ibid., 160.

25. Machiavelli's translator was Father Anton Rafa'el Zakhur, one of the first of Muhammad Ali's translators. Castera was translated by Jacoraki Argyropoulo. See Bianchi's "Catalogue général des livres Arabes, Persans et Turcs imprimés à Boulac en Egypte depuis l'introduction de l'imprimerie dans ce pays," in *Nouveau Journal Asiatique* II (Paris, 1843), 24–60; and al-Shayyal, *Tarikh al-tarjama*, 160–185.

26. The translator of this book was Muhammad Moustafa al-Baya`a, who after graduating from *Dar al-Alsan* worked as an editor of foreign translations. (Al-Tahtawi himself corrected the translation before it was published in Bulaq in 1841.) One of the first men to graduate from Dar al-Alsan, Khalifa Afandi Mahmoud, who later served as head of the Bureau of Translations, translated a number of histories from French to Arabic. See Salah Majdi, *Hilyat al-zaman*

bi-manaqib khadim al-watan. Sirat rifa`a rafa`i al-tahtawi, ed. Jamal al-Din al-Shayyal (Cairo: Wizarat al-Thaqafa wa-l-Irshad al-Qawmi, 1958), 43–54.

27. According to Crabbs, Muhammad Ali intended to have an encyclopedic history of his reign written, as well as memoirs. Apparently he was too distracted by other projects to have either work finished. See Crabbs, *The Writing of History*, 68.

28. On the relationship between behavior, progress, and the ultimate construction of ethnicity, see John Camaroff and Jean L. Camaroff, "Of Toteism and Ethnicity" in *Ethnography and the Historical Imagination* (Boulder: Westview Press, 1992).

29. The full title of the text is *Aperçu historique sur les moeurs et coutumes des nations: Contenant le tableau comparé chez les divers peuples anciens et modernes, des usages et des cérémonies concernant l'habitation, la nourriture, l'habillement, les marriages, les funérailles, les jeux, les fêtes, les guerres, les superstitions, les castes, etc* (Paris, 1826). In Arabic, the title was rendered *Qala'id al-mafakhir fi gharib `awa'id al-awa'il wa-l-awakhir*. On al-Tahtawi as author of his own geographies, see Eve M. Troutt Powell, "From Odyssey to Empire: Mapping Sudan Through Egyptian Literature in the Mid-19th Century," *International Journal of Middle East Studies* 31 (1999): 401–427.

30. *Programmes de l'enseignement primaire, et de l'enseignement secondaire, approuvés par arrété ministériel No. 849, au date du 16 September 1901*. Date and place of publication not available.

31. Rifa`a Rafa`i Al-Tahtawi, *Qala'id al-mafakhir fi gharib `awa'id al-awa'il wa-l-awakhir* (Bulaq, 1833). All translations, unless otherwise mentioned, are my own.

32. In Arabic, *bilad al-kufar*. I imagine that this must have been a concession al-Tahtawi made to French: he makes reference to cleanliness and its place both in Islamic law and "that of Moses," using all the honorific phrases traditionally used when mentioning their names, something Depping would not have done. This might also be a case of al-Tahtawi inserting himself as "author."

33. Ibid., 20–21.

34. Cited in Anouar Louca, *Voyageurs et écrivains Egyptiens en France au vingtiéme siécle* (Paris: Didier Press, 1970), 62. "Si les Franais s'adonnent aux sciences...s'ils jouissent d'un vie économique prospére et cultivent des habitudes et des moeurs de plus en plus raffinées, c'est gréce a un régime politique favorable."

35. Ibid., 104–105.

36. Geography as the very concrete practice of studying the surface of Egypt was also crucial to many of Muhammad Ali's modernization programs, such as the construction of railroads, telegraphs, irrigation canals. Here I am interested in geography as the more abstract practice of ordering the universe. As Winichakul has suggested, this "abstract" geography becomes more concrete when maps are used to give boundaries to nations once they have been "placed."

37. Dar al-Watha'iq al-Qawmiyya (Egyptian National Archives), `Ahd Ismail collection; Ta`lim, Box 12; anon., "Enseignement de la géographie en Egypte," *Buletin de la societé de géographie a Paris*; Deuxiéme série, 3 (1835).

38. *Programmes de l'enseignement primaire.* Al-Tahtawi also compiled the works of a number of European geographers, to which he added a lexicon of useful terms. The text was called *al-Ta`ribat al-safiyya li-murid al-jeographiyya*, and was published by Matba`at al-Amiriyya in Bulaq, 1834.

39. I have been unable to locate a copy of al-Tahtawi's translation, and have therefore had to rely on an English version.

40. The full title of Conrad Malte-Brun's text is a *System of Universal Geography, Containing a Description of all the Empires, Kingdoms, States and Provinces in the Known World, Being a System of Universal Geography or a Description of All the Parts of the World on a New Plan, According to the Great Natural Divisions of the Globe, Accompanied With Analytical, Synoptical and Elementary Tables.* In French, the book's complete title is *Précis de la géographie universelle; ou description de toutes les parties du monde sur un plan nouveau* (Brussels, 1829). The following passages were all taken from the English version, James Percival, trans. (Boston, 1834). Al-Tahtawi's Arabic translation of Malte-Brun first appeared in 1838.

41. Malte-Brun, *System of Universal Geography*, Preface, Vol. I, 3.

42. Patrick Wolfe "History of Imperialism: A Century of Theory, from Marx to Postcolonialism," *The American Historical Review* 102, 2 (1997): 388–420.

43. Malte-Brun, *System of Universal Geography*, 1–2.

44. Ibid., Preface, 4. My emphasis.

45. Heyworth-Dunne, *An Introduction to the History of Education in Modern Egypt* (London: Frank Cass, 1968), 104–105.

46. The second book published by Bulaq Press was an Italian–Arabic dictionary, in 1822.

47. According to Alain Silvera, most of the records from the early missions were destroyed in a fire in the Citadel, which formally housed the state's archives, in 1830. See his "The First Egyptian Student Mission to France under Muhammad Ali," in *Modern Egypt: Studies in Politics and Society*, Elie Kedourie and Sylvia Haim, eds. (London: Frank Kass, 1980), 1–22.

48. Ibid., 8–9.

49. See Rifa`a Rif`i al-Tahtawi *Takhlis al-ibriz fi-talkhis baris* in *al-A`mal al-kamila*, Muhammad `Imara, ed. (Beirut: al-Mu'assa al-`Arabiyya li-l-Dirasa wa-l-Nashr, 1973), Vol. 2, *al-Siyassa wa-l-wataniyya wa-l-tarbiyya*, 189.

50. *Evening Entertainment; or, Delineations of the Manners and Customs of Various Nations, Interspersed with Geographical Notices, Historical and Biographical Anecdotes and Descriptions in Natural History* (Philadelphia: David Hogan, 1917). According to Timothy Mitchell, the book's leitmotif was indolence, as it was opposed to productivity, and as it was found in the world's less civilized peoples. I think Mitchell is correct in noting that the texts gave specific, prescriptive formulas for wiping out indolence, and that making more productive Egyptians was part of the state's general plan for reform. Mitchell overlooks the book's other agendas, however, not the least of which was this process of assessing modernity through morals and behavior. See Mitchell, *Colonising Egypt*, 104–108.

51. There are numerous biographies on al-Tahtawi. The best discussions of his career in service to the state, as translator, educator, and administrator, are found in al-Shayyal, *Tarikh al-tarjama*; Majdi, *Hilyat*; and Gilbert Delanoue, *Moralistes et Politiques Musulmans dans l'Egypte du XIXéme siécle*, Vol. II (Cairo: Institut Français d'Archéologie Orientale du Caire, 1982), which contains a considerable bibliography on al-Tahtawi and his various endeavors.

52. The text was also translated into Turkish in 1839. See Anouar Lourca's translation of the text into French, *L'Or de Paris* (Paris: Sindbad Press, 1988).

53. Al-Tahtawi, *al-A`mal al-kamila*, Vol. 2, 4.

54. Delanoue, *Moralistes et Politiques*, 388.

55. Allen argues that *Takhlis al-ibriz* was the first in a series of works in which Arab writers recorded their impressions of Europe. He claims that the subject of the "grand tour" of Europe later served as the trope for a series of novels about the relationship between the East and the West that have appeared over the course of the twentieth century, and for autobiographies such as Taha Hussein's famous *al-Ayyam*. See Roger Allen *The Arabic Novel: An Historical and Critical Introduction* (Syracuse, NY: Syracuse University Press, 1982).

56. Al-Tahtawi mentions his translations of Depping's work in the second chapter of his third *maqala*, "Ahl Paris," or "The People of Paris" in which a number of institutions and customs are described. See al-Tahtawi, *al-A`mal al-kamila*, 75.

57. Al-Tahtawi admired French architecture and craftsmanship, but thought that the materials with which Parisian homes were built were generally bad, especially compared with Cairo. See *al-A`mal al-kamila*, Vol. 2, 107–109.

58. Ibid., 107.

59. Ibid., 108.

60. Ibid., 108.

61. Ibid., 108. The prevalence of literacy in France was noticed by numerous Egyptian travelers. Ahmed Zaki, journeyed to Paris in 1892 and 1900. In his accounts of his journeys there he noted that even drivers knew how to read. The existence of numerous libraries in Paris was proof, to him, of the extent to which the French were literate. See Louca, *Voyageurs*, 212.

62. Ibid., 107–110.

63. Ibid., 113–114.

64. Ibid., 132.

65. In Arabic: *al-Dafatir al-sanawiyya, al-taqwimat al-jadida, wa-l-zijaat al-musahaha*.

66. Ibid., 172.

67. Marsot, *Egypt in the Reign of Muhammad Ali*, chapter seven.

68. Cited in André Raymond, *Cairo*, Willard Wood, trans. (Cambridge MA: Harvard University Press, 2000), 301–302. See also Desmond Steward, *Great Cairo: Mother of the World* (Cairo: The American University in Cairo Press, 1996).

69. Janet Abu-Lughod, *Cairo: 1001 Years of the City Victorious* (Princeton: Princeton University Press, 1971), 94.
70. Ibid., 95 and 106.
71. Nihal Tamraz, *Nineteenth-Century Houses and Palaces* (Cairo: The American University in Cairo Press, 1998), 1.
72. Ibid., 26.
73. Hunter, *Egypt Under the Khedives*, 100.
74. Ibid., 99–110.
75. Toledano, "The Long Nineteenth Century," 274–275.
76. See al-Shayyal, *Tarikh al-tarjama*, part three.
77. Edward Said, *Orientalism* (New York: Vintage Press, 1994), 49–73.
78. Khaled Fahmy, *All the Pasha's Men: Mehmet Ali, his Army, and the Making of Modern Egypt* (Cambridge: Cambridge University Press, 1997).

CHAPTER 2

Modernity, National Identity, and Consumerism: Visions of the Egyptian Home, 1805–1922

Mona Russell

Although in modern Western society, we tend to equate the home with the private sphere, as this volume demonstrates, the home is connected to state and society in ways that vary over time and place.[1] In her seminal study of the antebellum American south, Elizabeth Fox-Genovese further argues:

> The household should...be seen as a symptom or manifestation of specific social relations...[P]olitical and legal institutions, the complexity, nature, and availability of markets, and the social relations of production...shape the ways in which households develop. Households may legitimately be viewed as the primary mediating units between the individual and society, as genuinely pivotal institutions, but they must be subjected to a scrupulous analysis of their specific character and activities in interaction with different social systems.[2]

Historically, elite Ottoman households united both the public and the private spheres. They were both the locus of power and the residence of the ruler and lesser officials, and such households were connected to one another by chains of command, blood relationships, marriage, and the exchange of slaves.[3] In Egypt Mamluk households served the same function both before and after the arrival of the Ottomans. André Raymond characterizes the aristocratic Mamluk household as both a symbol of conspicuous consumption and a seat of power. He describes it as a multigenerational, supporting as many as 200 people, including the owner, his family, slaves, and retainers. It was both a "manifestation of the power of the proprietor" and an imposing fortress built and decorated in a "sumptuous style."[4]

In this chapter, I will argue that the erosion of this architectural style and the centralization of power in a state, symbolized by the urban bourgeois household, helped to create both a modern Egypt and an Egyptian national identity. In particular, I will be discussing the emergence of this home, focusing on Cairo at the turn of the nineteenth century. Furthermore, I will be dealing in visions of the ideal home, as it existed in prescriptive literature, textbooks, and advertisements, not necessarily the average Cairene home. Finally, I will be linking these topics to the emergence of the New Woman in Egypt.

The structural and bureaucratic changes that took place during and after the reign of Muhammad Ali created a state that encouraged the participation of more households in national life and which, as Lisa Pollard argued in chapter 1, vigorously promoted how those households should be constituted. Greater participation in public life was not synonymous with democratization. It was an intrusive, "top-heavy" state that created a modern Egypt. Muhammad Ali's educational, bureaucratic, and military reforms allowed a new bourgeoisie of technocrats, functionaries, officers, and professionals to advance socially and economically over the course of the nineteenth century. In the countryside, a class of local leaders advanced as well; and while they remained there, they sent their sons to the new schools in Cairo and other urban centers. Further, as demonstrated in chapter 1, these changes had implications for where the elite chose to live and how they chose to constitute their homes.[5]

Some of these reforms waned under the leadership of Abbas (r. 1848–1854) and Said (r. 1854–1863); however, they continued to expand Egypt's transportation and communication systems. Aided by the boom in cotton during the Civil War in the United States, Ismail (r. 1863–1879) sought to refocus on order and education. His great hope was to turn portions of Egypt, in general, and Cairo, in particular, into extensions of Europe, in order to encourage foreign settlement and investment. Although Ismail's plans landed in Egypt in a vortex of spiraling debts, he effected changes in the country's education and infrastructure that would continue the process started by his grandfather. These changes meant that a whole class of individuals began to shape the nature of the Egyptian state, rather than just the individual occupying the citadel. While the state laid the infrastructure and set the incentives for change in the first half of the nineteenth century, it was choices adopted by individuals and their application in the late nineteenth and early twentieth century that helped to forge the new identity.

There existed a synergistic interplay between state plans and incentives, the role of a self-interested elite, and individual choice. Examining state sponsorship of certain neighborhoods using the cases of Azbakiyya, Abbasiyya, and Heliopolis are illustrative here. During the occupation (1798–1801) the French established their headquarters in Azbakiyya. They built new streets and created a bustling service sector of taverns, cafes, restaurants, and dance halls, a trend that Muhammad Ali continued.[6] Ismail earmarked Azbakiyya as the hub for his new Cairo, and he renovated its park and gardens, sparing no expense.[7] He even hired Jean-Pierre Barillet-Deschamps, the former chief gardener to the city of Paris, to do the landscaping.[8] An 1896 guidebook to Egypt is testimony to his efforts:

> The gardens are well kept, and the trees have grown quickly and give good shade. It is a very favorite place of resort, especially when an English or Egyptian military band performs in the evening. The cost of making the gardens was considerable, and the inhabitants and visitors to Cairo owe a debt of gratitude to Ismail Pasha for forming this tasteful pleasure-ground.[9]

Creating parks and gardens was a priority for state planners, who believed that they would bring harmony, order, and hygiene to the harsh, urban environment.[10] Such planners were often first generation city-dwellers, who longed for a reminder of the countryside.

Just west of Azbakiyya, Ismail created a new subdivision as his own namesake. His new quarter would have streets laid out in grid fashion with gas lighting, and its homes would be served by running water. To expedite settlement he offered free land to anyone who would build a suitable— meaning worth at least £E 2,000—home.[11] These changes served the elite and foreigners. Ali Mubarak reports that by the end of Ismail's reign only 200 homes (mansions) were built.[12] In contrast, the experience of Abbasiyya filtered down to the middle class and demonstrated the interchange between state policy and individual initiative.

Abbas, Muhammad Ali's first successor, attempted to facilitate the development of his namesake by building a palace and military barracks there, as well as by compelling the elite to build their homes there. Nonetheless, his successor Said thwarted the neighborhood's development by relocating the military barracks to Qasr al-Nil.[13] Said's efforts were so successful that an

1867 guidebook describes Abbasiyya as follows:

> The town founded by the late Abbas Pasha . . . is a miserable memorial of the wish on the part of its founder to ennoble his name, without considering whether the object was useful, or the monument likely to endure.[14]

The neighborhood's turnaround began during the reign of Tawfiq (r. 1879–1892), who resided at al-Qubba palace just beyond Abbasiyya, once again encouraging elite settlement.[15] Nevertheless, the huge push to settle there did not come until the arrival of mass transit in the closing years of the nineteenth century. The tramway, which ran along Sharia al-Abbasiyya, created two distinct sectors: an Eastern elite district and a Western middle class district.[16] Thus, the state can limit and create opportunities for its citizens by building roads, extending mass transit, and encouraging development. At the same time, development of neighborhoods is contingent on the choices made by individuals.

The development of Heliopolis, or New Egypt (*Misr al-Jadida*), was also intricately tied to the development of the tramway. Not surprisingly, the owner of the first tramline concession was also the founder of this district. Within a few years time, land values had, in some cases, increased four-fold. Speculators rushed to invest in land at the end of the line for casinos and resorts. By 1905 a groups of financiers came together to develop the area further. Within four years Abbasiyya and Heliopolis were connected by a modern road and an electric tram, and the new district contained eighty villas and shops, a hotel, and a casino. Although foreign interests ran high in Heliopolis, it could not be defined as a European town. Its European population was considerable, but the percentage of Europeans did not exceed twenty percent.[17] Nevertheless, it remained a privileged spot, one of the two tram lines that "white" (foreign) people could ride.[18] It received more than its fair share of attention as a cartoon from 1921 in *al-Lata'if al-musawwara* indicates. It depicts the Minister of Transportation booting a female Helwan down a flight of steps and uttering the triple divorce oath, while a coy Heliopolis watches with delight.[19]

The redistribution of power in Egypt, the expansion of new neighborhoods, the growth of the urban middle class, and the spread of new architectural styles allowed greater numbers of Egyptians to utilize their homes as a site for conspicuous consumption. The expanded class of bureaucrats, officers, teachers, technicians, and a growing class of urban professionals sought new housing in the new neighborhoods.[20] Furthermore, the wider adoption of the home as a site for and a symbol of conspicuous consumption demonstrated the spread of a national identity among an urban middle class. In other words, more urban Egyptians felt as though they had a stake in the system, and their home was a symbol and manifestation of that stake. This connection was made clear by state-produced texts (as discussed by Pollard in chapter 1), government and private school curricula (see later), and by the mainstream and women's press.

The rise of a women's press, as well as columns in the mainstream press for women, led to a whole genre of prescriptive literature on how to organize the new home. Women's columns and magazines were "showcases" for the changes that had taken place in Egypt over the course of the nineteenth century, as well as "sites of cultural production, generating class, gender, and national identities while selling goods and fashionable lifestyles."[21] According to this genre, the first prerequisite of the modern home was its location in a suitable neighborhood. One of the most remarkable characteristics of the urban history of Cairo in the nineteenth and twentieth centuries is its expansion to the north and northeast, made possible by the spread of mass transit and the building of new bridges.[22] The process of urban renewal begun under Ismail created the open squares and thoroughfares that served as the building blocks for Egypt's mass transit system, which developed in the last years of the nineteenth century. According to Abu-Lughod, between 1897 and 1917 Cairo was able to triple its area without increasing the time that it takes to travel from its most remote areas to the center of town.[23]

With new districts such as the elite suburbs of Heliopolis and East Abbasiyya, and the more middle class suburbs of West Abbasiyya, Faggala, and Zahir, choosing the right neighborhood became all the more important. Turn-of-the-century journals defined the proper neighborhood as sanitary; well-ventilated; equipped with utilities, hospitals, and schools; and filled with good neighbors.[24] The state did its part to foster these neighborhoods by building better streets, providing lighting and public utilities, and imposing order by numbering homes. An anecdote from Muhammad al-Muwaylihi's *Hadith `isa ibn hisham* is useful here. The book is a Rip van Winklesque tale of a pasha from the age of Muhammad Ali awakening in the last years of the nineteenth century. When a writer named Isa encounters the revived Pasha, the latter cannot believe that Isa does not know who he is, nor does he know the location of his home. The Pasha lived in the transitional period between the great *amiral* households and the new home, and at that time homes were still known by the owner's family. Since the Pasha's family was a prestigious one, he assumed that anyone would know where his home was located. Isa informs the Pasha that houses are no longer identified by their owners, but rather "[they are known] by their streets, alleys, and addresses."[25] In other words prominence would come not from the owner, but from the neighborhood in which he lived. State planners believed that greater order would come from changes such as widening roads and numbering homes.[26] These changes first took place in the most elite neighborhoods in the city, and then gradually moved to the suburbs.[27]

The new home required more than just the proper neighborhood. Prescriptive literature cited a second category of prerequisites that included suitable building materials, an architectural form adhering to new notions of public and private, and appropriate interior decoration. No longer would homes be imposing, intimidating structures on the outside with inward-looking courts and gardens. Again, a return to Muwaylihi's *Hadith `isa ibn hisham* is useful here. The Pasha's grandson lives in a mansion surrounded by a garden, rather than the reverse.[28] Instead of employing the traditional *mashrabiyya*, these new homes utilized high walls and gates for privacy. In these changes the home lost the *cupola* and other features that made greater sense for Cairene weather. Writing in the 1870s after a tenure as American Consul, Edwin de Leon remarked that the architecture of the new home was "replacing the most picturesque" and "comfortable" with the "ugliest."[29] The adoption of such habits was apparent by the 1890s. In an article from *al-Muqtataf*, the author points out the foolishness of taking one's overcoat off in Egypt where the temperature inside the [new] home is as cold as, if not colder, than the outside.[30] Where the old architectural style had promoted circulation of air, the new style tended to retain heat in the summer and cold in the winter.

Previously, the location, style, and materials of the Mamluk home had been important factors; however, the idea of dividing the home into rooms with fixed functions and furnishings was new. Aside from rooms set aside for visitors [male and female], as well as the kitchen and bath, rooms in a traditional, Mamluk-style home had great permeability and flexibility. Any room could be used for sitting, sleeping, or dining by simply arranging or rearranging cushions, mattresses, and trays.[31] At the minimum, the new home necessitated a foyer, kitchen, reception area, sitting room, and bedrooms. A house that contained a library, or even just a desk for the wife, was considered better since it set an example for the children. Furnishings for the home were to be both beautiful and functional, without jeopardizing the household budget. Furthermore, the house was to be treated like a growing, living entity that required appropriate maintenance. Rooms had to be cleaned, upgraded, and decorated regularly.[32]

The third characteristic of the new home was that its happiness rested upon the shoulders of the woman. Although not explicitly labeled as such, this individual was the New Woman. Many inside and outside Egypt associate the notion of the New Woman with Qasim Amin and the 1900 publication of his book by the same title (as well as his previous work *Tahrir al-mar'a*); however, Amin was merely one current in a much larger stream of writings.[33] Examining the new publications section of journals for example, *al-Hilal* and *al-Manar* indicates that such writings were present both before and after the publications of Qasim Amin. As Leila Ahmed has persuasively argued, Amin's New Woman was neither new nor was she particularly Egyptian.[34] Instead he created a class-based critique of Egyptian society that generated a flurry of debate in the press.

The New Woman of the mainstream and women's press was someone who embodied the traditional values of her female forbears, yet superceded them in her ability to run the home, administer its finances, educate her children, and serve as her husband's partner in life. Perhaps one might question how "new" this role was for women. It is clear that middle-class and elite women were not the slovenly *odalisques* portrayed by Western travelers and artists. Nevertheless, evidence as to what role they played in buying for the home has been extrapolated backward from the nineteenth century.[35] Even if the role were not new for elite women, growing numbers of middle-class women were just starting to take on these duties.[36] While some women may have been educated in their homes and provided some instruction for their children, there was certainly no uniformity of practice.[37] As for care of the children, the use of *dadas*, eunuchs, and wet nurses was quite common. Elite women supervised the care of their children; however, they did not necessarily participate actively. Finally, with respect to the partnership of marriage, Islam sanctioned roles for men and women; however, historically as in the West, there was little notion of compassionate marriage.[38] Furthermore the existence of polygamy and concubinage further distanced the elite from this concept. The large multigenerational households of the Mamluk age were far different from the new home and family, which although not explicitly so, was implicitly nuclear.[39]

The New Woman was to provide a hygienic, attractive sanctuary from the world outside its walls. In the words of a male contributor to a woman's magazine in 1901, "the Eastern woman is raised for married life, that is to be a wife, a mother, a household manager, and one who rears the children."[40] In order to uphold these roles, she had to be the general administrator and purchasing agent for her home. These duties included being a meticulous record-keeper.[41] Although the mistress of the new home was likely to have servants, she was responsible for planning, supervising, and monitoring their work.[42] She was to take advantage of all scientific and technological advances to keep her home in tip-top shape.[43] During eras of inflation, economic crisis, and war, these skills became even more important.[44] Obviously notions of thrift and practicality played a more significant role in the lives of middle-class women than those of the elite.

The wife was a partner to her husband, and their roles were complementary. She, with her gentle nature, would serve as a mental and emotional support to her husband and family. In return, he would labor outside the home, providing its material support. She depended on her husband, who in turn was strengthened by her dependency. Nevertheless, she would be the strongest force in her husband's life, creating a "separate, but equal" partnership. Since one spent most of his/her life married, the entire happiness of the material world depended upon the success of this partnership.[45] In addition to maintaining the home physically and providing emotional support to her husband, the woman was responsible for the other inhabitants of the home, namely her children. Egyptian mothers had been caring for children since the dawn of civilization, but what was new was the role as mother-educator—one that neither uneducated women nor women who left their children in the care of others could partake.[46] Cast as mother-educator, the woman could pursue a nationalistic and quasi-political role without upsetting the patriarchal balance. Furthermore, this ideal coincided with the British occupiers' desire to stem the tide of feminism at home and abroad.[47]

In caring for the new home and its inhabitants, the New Woman was caring for an entity that was symbolic of the family within it and the community surrounding it. Herein lies the link between the home, consumption, and national identity. The home was both a small kingdom and the foundation of a larger entity, whose past, present, and future condition could be judged by the state of its individual kingdoms. Indeed, the nation was a collection of homes, therefore reform of the nation was tied to reform of the home.

The linkage of consumption and identity was not unique to prescriptive literature from the women's press and women's columns. Perhaps even more significant and of longer-lasting importance was the advice given in textbooks for the growing numbers of schools for girls.[48] Education was a key factor in creating the urban bourgeoisie, and it remained an important component of the nationalist agenda. As discussed in chapter 1, Rifa`a al-Tahtawi was an early product of Muhammad Ali's reforms, and he later became a key decision-maker in designing Egypt's educational structure. His writings represent an interesting blend of Azhari and post-enlightenment thinking.

According to al-Tahtawi, education should have religion as its foundation, but it should also include political education so that an individual can understand his/her role in his/her nation. Furthermore, education should extend to girls as well as to boys. His key argument was that educated spouses make better partners for educated men and that reformed homes and families lead to a reformed nation.[49] His experiences in Paris shaped his future career in Egypt as he translated or supervised the translation of numerous works. As discussed by Pollard in chapter 1, he wrote an account of his stay in Paris, in which he provided a meticulous description of the Parisian home and its upkeep.

Turn-of-the-century textbooks mirror his descriptions with great precision. Homes with proper divisions, floors that are waxed regularly, decorated fireplace mantels, dens filled with documents and books, and pianos all make their way into home economics textbooks. Men and women, Egyptians and foreigners, believed that home economics was the most significant subject for girls; however, men tended to push the practical aspects, for example, the basics of cooking and cleaning, while women advocated its theoretical and philosophic aspects, finance, and child-rearing. The amount of theoretical versus practical home economics depended upon whether one attended a higher or lower track government school, missionary school, private Muslim school, or one of the many schools belonging to the minority communities of Egypt, with variation occurring even within each division.[50]

The lines between hygiene, morals, home economics, and civics were often blurred in turn-of-the-century textbooks, and such subjects were often combined in the lower educational track. The creation of a strong, healthy nation necessitated strong, healthy individuals who created strong, healthy homes. This process involved the judicious selection of certain Western values and commodities. Nevertheless, textbook authors emphasized that these practices should be imbued with Eastern values.[51] For example, Antun al-Gamayyil wrote the following in his introduction to a 1916 textbook: "there are many books on female upbringing from all nations," and that these books have a profound influence on "national life," but they are not suitable for "our Eastern life and our national customs" [hiyatna al-sharqiyya wa `adatna al-qawmiyya].[52]

The linkage of hygiene, health, home economics, and national reform were repeated time again. Chapters on cleanliness of body were juxtaposed with those on behavior, which were then followed by chapters on maintaining the home physically, spiritually, and emotionally. The introductory and concluding sections of these chapters frequently related the above subjects to the health of the nation. Home economics textbooks generally took students on a tour of the house with the proper divisions. The ordering of the individual home was reminiscent of the restructuring of cities since the time of Ismail through the creation of new streets, parks, and urban amenities, as well as police, fire, and traffic brigades. Similarly, women were to impose the same order, regulation, and divisions in their homes, while providing an aesthetic appeal. The individual home was intricately connected to the surrounding society. Al-Gamayyil points out that men with dirty, noisy homes find it necessary to escape, seeking refuge in taverns and gambling parlors, where they lose their health, wealth, and morals.[53]

Within the properly divided home, having a place for books, whether in a den or another room of the home, was extremely important. The home was the child's first school and the site of continuing education after school hours. One author, reminiscent of the al-Tahtawi selection (in chapter 1, p. 24), even allotted money for books and journals in his sample budget for the home citing moral, intellectual, and nationalistic reasons:

> The individual cannot dispense with books and journals. The best homes are those which are adorned with a bookcase or bookcases, as it is better that they are preserved for the individual and his children. Books are a good form of entertainment and the best type of solace, since many benefits and much knowledge is gained from them; and they fill [one's] spare time with a useful activity. In books, [one finds] that which is necessary for every [living] creature from history, literature, stories, amusements, notable deeds, happiness, wisdom, and advice...
>
> ... the individual cannot dispense with newspapers from which [s]he reads the news of his [or her] nation [*watanihi*] and the events, happenings, and politics which take place outside his [or her] country [*biladihi*].[54]

Furthermore, it was the woman's job to maintain the family library, arranging the family holdings and even creating a card catalog. She was also to keep a "golden family book" with pictures, souvenirs, and handwritten documentation of significant family events.[55] As will be seen in chapter 3, Mickelwright argues that such collections are symbolic of both modernity of identity construction.

Since the home was the child's first school, his mother was naturally his first teacher, and textbooks reinforced the vision of mother-educator touted by the mainstream and women's press. The model was implicitly, and sometimes quite explicitly, that of the nuclear family with complementary roles played by fathers and mothers. Al-Gamayyil writes, "the man wants to be the lord who is unchallenged in everything...." However, he also points out that the woman is usually able to win the man over to her opinion.[56] These sentiments are echoed in other textbooks as well:

> [T]he wife must submit to her husband and give him legally sanctioned obedience, in service to him and his sons, concerning herself with the

arrangement of the house, administration of meals, cleaning, and economy, as well as the upbringing of [their] sons and daughters.[57]

Nevertheless, this author too put restraints on the senior partner, informing the wife that it is her husband's duty to provide material support and that he should not waste his family's money or time in amusements outside the home [drinking and gambling].[58] Both authors see the home as a microcosm of the community, and thus carrying out familial roles served both the family and the nation.[59]

As was the case in prescriptive literature of the press, girls learned about their somewhat contradictory roles as general purchasing agent and financial manager of the home. On the one hand, they were encouraged to adopt lavish Western styles of consumption, while on the other they received repeated admonitions against wasteful spending. Consider the following description of a foyer:

It is the entrance to the place of residence, and . . . it is [like] the saying "take [choose] the book from its title." It is the first area to welcome all who enter. Thus, it is necessary . . . that it not be devoid of welcoming adornment. . . . In the presence of darkness or in a foyer whose light is absent, place a mirror appropriately so that bright light is gathered in its reflection. Likewise, it [the foyer] is beautified by placing some colorful Japanese crafts, relics produced by previous peoples, antique weaponry, the horns or skins of predatory animals, or any similarly rare objects along its walls.

The good taste of the lady of the house is [evident] in [her] sound judgement in correctly placing things . . . the foyer is among the [most] beneficial places . . . because the visitor encounters everything which he needs from taking off his overcoat to leaving his walking stick, etc. It is recommended that it have comfortable seats; however, it should not be over-decorated with too much furniture or too many objects. It can be traditionally decorated with cushions, furnishings, and chairs, as long as they are elegantly situated. If the foyer is in a modern dwelling, then it will be lit by electricity, whereby a sealed electric current flows through a chandelier, candelabra, or lamp.

A tray should be set on a table so that visitors' cards can be placed in it, as well as simple writing implements for writing notes or jotting down words [thoughts]. As for the skirting [trim] of the foyer . . . it should be covered with white wood which [has been] coated with varnish or paint close in color to the furnishings. The remainder of the wall should be painted with an oil based paint or something similar to that. The pictures which hang on its walls should not be expensive. Similarly, the floor should be covered with a darkly-colored carpet which is inexpensive, and the curtains should be devoid of decoration or embellishment. No screen, or anything like that which is usually placed at the door of the harem, should be placed there; [because] it is against the conditions of health, and it is an out-moded form of decoration.[60]

The author, although allowing for some traditional items, clearly encourages Western forms of consumption and behavior. He exhorts the lady of the house to purchase things like Asian imports, antiques, and a tray for visitors' cards.[61] In his discussion of the salon (the room in which guests are received), Mikhail informs the reader that photographs of the family be placed there since there might be guests unfamiliar with the family. As well he recommends placing a portrait of the lord and/or lady of the house, taken by a reputable photographer or drawn by a well-known artist. Traditionally, displaying images of people was uncommon, let alone images of one's wife and children in a location under quasi-public scrutiny. Similarly, as was customary in Europe, Mikhail encourages decorating with pictures of the royal family, famous ministers, and or other great personages.[62] The careful, "museum-like" placement of such socially constructed objects was a clear statement of the family's composition, standing, and identity. At the same time, Mikhail discourages spending money on traditional items, for example, Oriental carpets and *mashrabiyya* screens. He even goes so far as to accuse the latter of being a menace to one's health. In another textbook, he implicitly supports the intermingling of the sexes in social situations by explaining that women should only shake hands with other women.[63]

The author of this particular textbook, Francis Mikhail, produced several for girls; and unlike many textbooks of the time, his contained photographic representations, which amplified his message of consumption. His meticulous descriptions were followed by pictures of Louis XIV-style furniture, Western lighting fixtures, English tea services, pianos, billiard tables, and fireplaces. Mikhail also frequently included lists of necessary items, for example kitchen implements and their prices, nor was he beyond endorsing brands, for example, Singer sewing machines. At the same time, however, he would also include chapters on economy and thrift, to remind young women of how to balance wants, needs, and resources.[64] Indeed, the purpose of the sewing machine was so that one could make her family's clothes, as well as recycle them.[65]

Once a young woman graduated from school, married, and began keeping her own home she could continue her education through prescriptive literature in the mainstream and women's press, as well as by advertising in both presses. While advertising at the turn-of-the-century was not as widespread as in the West, it was a growing field. Previously publishers had relied heavily on subventions and/or subscriptions, but by the early years of the twentieth century remarkable growth took place, particularly as government support for the press declined.[66] As early as 1897, the editors of *al-Hilal* penned an article touting the benefits of advertising, and they promoted their reasonable rates and capabilities as advertisers, claiming to reach "thousands of readers fivefold."[67] Alexandra De Avierino, editor of the women's magazine *Anis al-jalis*, routinely dedicated 25–30 percent of her publication to advertising and wrote an article entitled, "The Benefits of the Overlooked," in which she argued that advertising benefited the nation by fostering an open exchange of goods and ideas, since more publications could proliferate.[68] Products for the new home and its inhabitants routinely appeared in the pages of the Egyptian press.[69]

Between the late nineteenth century and the First World War, advertising demonstrated a fixation with modernity and progress, and companies often boasted about the foreign origins of their goods. In the period between the First World War and partial independence in 1922, advertisements reflected a trend toward nationalism and Egyptian identity, even with respect to foreign products. Companies advertised utilizing statements for example, goods produced by "the hands of Egyptian workers," or "national craftsmanship" produced by "the hands of citizens."[70] Many firms espoused a nationalist image to gain popular support, despite ties to foreign capital.[71] Even Palmolive soap was not beyond capitalizing on nationalist sentiment emanating from the post-1919 era. A 1922 advertisement depicts a "Cleopatra-type" figure gazing contentedly as two dark-skinned slave men mix some sort of concoction in a large tureen. The text discusses how ancient Egyptians used olive and palm oils as natural resources for beauty, the same resources to which the modern Egyptian has access via Palmolive.[72] Lactagol, a nursing supplement available through a foreign agent, as well borrowed nationalist images and themes to sell its product. The central figures of this 1922 advertisement are two women carrying a banner. One of them is upper class as revealed by her stylish shoes, calf-length cloak, and light *yashmak*. The other woman is of more humble origins as seen by her slipper-type sandals, more modest cloak, and *burqu`*. The banner reads, "Oh nationalist mothers, among the most sacred of your duties is raising healthy children for the nation, use Lactagol."[73] The company clearly sought to create a linkage between consumption of their product, nationalist feeling in the wake of the 1919 Revolution, and the emergence of the New Woman in both her middle- and upper-class forms.

Advertisements from the turn of the century may provide information on what was available for sale during that time, fads, when products first attained mass distribution, and in the case of testimonial advertisements, with whom advertisers thought consumers could identify and trust. Nevertheless, they do not tell us how they were received by the individual reader.[74] One small clue comes from watching the development of advertising for specific products and by specific agencies, as well as noting which advertisements run for a short duration. For example the same company ran a series of advertisements for beauty products in 1892; however within a few years they disappeared. Some of the products, for example, a French weight loss medication, may not have had resonance with the population. Many Egyptians, men and women, probably would have found the before picture more appealing than the after one. Since these advertisements required response to a post office box, it was perhaps apparent that the products were not well received.[75] It is from this process that we can speculate about how certain products were received and which ones required more marketing, different strategies, or different target audiences. Furthermore, it is clear that not all Western products and services were desirable. Regardless, the growth of advertising and numbers of stores selling Western products (even if indigenously produced) indicates a greater adoption of Western goods.

Certainly it is worth noting the irony that adopting certain forms of Western consumption helped Egyptian women formulate a national identity and seek to participate in national life by creating the proper home. The 1919 Revolution highlighted this important role as demonstrated by the iconography, rhetoric, and active participation by women.[76] Nevertheless, to assume that this process took flight during the period before the First World War era, under the British occupation (1882), or even before, after the French occupation (1798–1801), is simplistic. The antecedents were already underway in the eighteenth century, as *dallalas* marketed French products to elite and middle-class women, and the French surveyed the Egyptian market, sending samples of textiles home to be copied.[77] This process continued over the course of the nineteenth century with the arrival of large numbers of southern Europeans and Levantine traders. These individuals brought new commodities and ideas to Cairo and Alexandria, serving as channels for new modes of consumption.[78]

The school system established by Muhammad Ali and revitalized by Ismail encouraged the adoption of new ideas, products, and services. As well the arrival of missionaries facilitated this process. Muhammad Ali realized that the state could not bear the burden of education alone, and he encouraged missionary societies by granting them land, buildings, and even free train transportation.[79] Under Said and Ismail, missionary education spread rapidly. Given the government's inability to provide education for all those desiring it, such schools succeeded. Schools for girls were particularly important for three reasons. First, despite the lip service that the British paid to improving women's condition by means of education, under Cromer (1882–1907) they were reluctant to provide the necessary funds to meet the demand for female education. Second, where government schools for boys were deemed preferable by Egyptians given the potential of government employment after graduation, such conditions made government girls schools less attractive for women of the upper classes. Third, male and female missionaries targeted the home as the logical starting point for development and reform. The approach was two-pronged, through girls' schools and through visits to Egyptian homes. Charles Watson of the American Mission explains as follows:

> This work among women is most valuable in breaking down pernicious social customs. It is an acknowledged fact, that in every country social and religious traditions strike their deepest roots into the life of womanhood. No country can free itself from the tyranny of such customs until the womanhood of that country is influenced. Harem work has had a remarkable influence along those lines in Egypt.[80]

Teaching the girls how to keep house and design their homes was extremely important for the Christianizing process. Minnehaha Finney describes her pride in her students' housekeeping skills saying that "compared to the homes where the girls have had no education there is much to admire."[81] While the rates of conversion remained low, the influence in education and housekeeping was of much more significance. By the turn of the century, the American Mission alone had 119 schools serving 8,000 students, and within another two decades there were more than 2,000 students in just Cairo with schools in 175 villages.[82] Girls' schools varied widely in size and quality from their flagship institution, the Cairo Girls' College, to no more than a room in a missionary's home overseen by his wife and an assistant.

In addition to foreigners and missionaries, as we have already seen, the royal family also served a conduit of change in modes of consumption. While Ismail may have created oases of European consumption in Cairo and other large cities of the delta, he also transformed his own home(s) into new models to be copied by the elite.[83] He built new palaces and renovated old ones, furnishing them with the most sumptuous Western styles. He imported a wide variety of meats, cheeses, fruits, and vegetables, in addition to a range of beers, wines, liquors, and tobacco for consumption in his home. Ismail sought the latest technology for his home, and he was particularly enamored of photography and related gadgets. He bought expensive equipment, photographs from around the world, and portraits of himself and his family.[84] The Khedive extended his avid consumerism to personal items for example, clothing, jewelry, and pharmaceuticals, for himself, his mother, the princesses, and his consorts. His interest in things European was not limited to just things, but also intellectual commodities, for instance, as education for his male and female children and subscriptions to dozens of European newspapers and journals.[85] Habits adopted by the royal family spread to high government officials by the late 1860s and more widely into the upper and upper-middle class by the turn of the century.[86]

This process did not go unchallenged. There were many critics of the new modes of consumption in articles and cartoons in the press, as well as in popular literature. The previously discussed *Hadith `isa ibn hisham* served as a vehicle for Muwaylihi to critique the new homes, neighborhoods, courts, professions, and leisure activities. Similarly, Abdullah Nadim used his journal *al-Ustadh* as a mechanism for criticizing the blind adoption of Western customs and manners. Both men felt that there were certain things to be admired and gained from the West, but that such things needed to be restrained by indigenous customs and values.

As was evident from such critiques, many Westernisms were adopted without respect to indigenous conditions, such as the previously discussed architecture or the extreme difficulty in cleaning the new home given the exigencies of living on the fringe of the desert. Mamluk-style closets, cabinets, cushions, mattresses, and trays were more easily cleaned and cleared of dust than the new styles of interior decoration. Still other things, such as European lavatory were adopted, or perhaps adapted is a better word, but not necessarily used in the prescribed manner. André Aciman's memoirs include an anecdote regarding his *Ladino* grandmother. He recalls having seen her, when he was a child, perch herself on top of the toilet with her bare feet on the rim. She explains that she can only use the bathroom in a Turkish manner. This incident took place in the 1950s, and thus surely the earlier transitional period was more difficult.[87]

Utilizing the home as the locus of change, bureaucrats in the centralized state, and individual citizens implementing free choice helped to create a modern Egypt and with it a national identity. Policies implemented by Muhammad Ali and his successors with respect to the expansion of the bureaucracy, education, municipal improvements, transportation, and communication created the infrastructure under which such developments could take place. Symbolically, this change came as power was diffused from an omnipotent household to a class of bureaucrats working for the state. This class was characterized by new modes of consumption with respect to education and in the creation of the new home, with the New Woman at its helm. An urban professional class developed alongside the bureaucratic class to help fulfill its expanded needs: banking, insurance, medicine, and commerce. Additionally, there arose a new economic sector to provide goods and services for the new home, including stores, pharmacies, restaurants, tailors, photographers, governesses, tutors, and schools. The state imposed its vision of the new home in state-sponsored curricula, and Egyptians responded to this vision through an ongoing dialogue in the press. Advertising, while providing little information with respect to how people actually shaped their homes, does help us to understand these changing dynamics.

While it should be emphasized that my topic here deals with the ideal vision of the home and not necessarily the homes belonging to the majority of Egyptians, I would like to conclude with an anecdote regarding the remarkable staying power of this vision. Louis XIV-style (or as foreigners refer to it, Louis-Faruq) furniture that was brought to Ismail's Abdin palace or as it existed in the pages of home economics textbooks from the turn of the century, came to represent the ideal salon to the point that most middle-class couples acquired such furniture as part of the wedding trousseau by the mid-twentieth century. Only in the decade after *infitah* did this style start to fade. When I was in Egypt in 1991, I overhead the remarks of a mother-in-law regarding her daughter-in-law's choice of furniture. Bear in mind that the mother-in-law viewed the young woman as not good enough for her son. In actuality they were just at slightly different points on a middle-class scale. The bride-to-be had chosen non-Louis-Faruq furniture, and her mother-in-law had this to say: "Her taste is so horrible, the colors are terrible, and the entire image is completely peasant-like [*baladi*]," meaning that one still needed the "right" furniture to have the proper bourgeois home. The further irony is the use of the term *baladi* to denigrate the young woman's taste, when the term quite literally means native or indigenous.[88]

Notes

1. In addition to its relationship with the state, the household per se is a dynamic concept that evolves over time. See, in particular, arguments made by Tania Forte in chapter 6 regarding the transformative nature of the household, using Deir al-Asad as a case study.

2. Elizabeth Fox-Genovese, *Within the Plantation Household* (Chapel Hill: University of North Carolina Press, 1988), 83.

3. For a discussion of this process as it relates to the Ottoman Empire, see Fatma Müge Goçek, *Rise of the Bourgeoisie, Demise of Empire* (New York: Oxford University Press, 1996), 28.

4. André Raymond, "Le Caire Sous Les Ottoman (1517–1798)," in *Palais et Maisons du Caire, Vol. II, Epoque Ottomane (XVI–XVIIIéme siécles)*, Bernard Maury, André Raymond, Jacques Revault, and Mona Zakariyya, eds. (Paris: Editions du Centre National de la Recherche Scientifique, 1983), 31.

5. One of the significant changes was that elite homes no longer required fortress-style residences and a place to garrison retainers. Over the course of the nineteenth century European states outlawed slavery both at home and in their far-flung empires. With the creation of new armies both in Egypt and in the Ottoman Empire, the need for male slaves subsided, but the trade in female slaves remained brisk.

6. Doris Behrens-Abouseif, *Azbakiyya & Its Environs from Azbak to Ismail, 1876–1879* (Cairo: L'Institut Fraçais D'Archeologie Orientale, 1985), 74–88.

7. Foreigners and Egyptians with close connections to the development project benefited from the rampant speculation that took place during the renovations, a process similar to the one that will be discussed by Ayşe Buğra in chapter 4 with respect to housing projects in Republican Turkey. The most expensive plots were those closest to the Shepheard's hotel, a location that was formerly the residence of Alfi Bey, then the headquarters for the French occupation, and later Muhammad Ali's Dar al-Alsan—a fascinating statement about the development of housing and real estate in Cairo. Nina Nelson, *Shepheard's Hotel* (London: Barrie and Rockliff, 1960), 8, 16, 25.

8. See Dar al-Watha'iq [DW], *Période Ismail* [PI], doss. 62/3 for materials related to the Azbakiyya renovation.

9. Mary Broderick et al., eds., *A Handbook for Travelers to Lower and Upper Egypt* (London, 1896), 327.

10. DW, *majlis al-wuzara'* [MW], *shirkat wa-jama'iyat* [SJ], box B4, series 224, Correspondence Relating to the Creation of a Public Garden in Alexandria, August 19, 1883–September 20, 1883; Correspondence Relating to the Establishment of a Public Park in Alexandria, September 1886–December 1889; Series 630, Note to the Council of Ministers from M. Fakhry Regarding Land around the Old Ismailieh Canal, November 20, 1920. As Timothy Mitchell has argued, efforts to tame and organize were part and parcel of the state's modernity project. See his *Colonising Egypt* (Cambridge: Cambridge University Press, 1988).

11. `Ali Mubarak, *al-Khitat al-tawfiqiyya al-jadida li-misr al-qahira wa-mudunha wa-biladha al-qadima wa-l-shahira* (Cairo: Dar al-Kutub, 1969), vol. 1, 210–211; Edwin De Leon, *The Khedive's Egypt* (London, 1882 [1877]), 32.

12. `Ali Mubarak is referring to the zone between Shubra and Ismailiyya. `Ali Mubarak, *al-Khitat al-tawfiqiyya*, vol. 1, 214.

13. `Ali Mubarak, *al-Khitat al-tawfiqiyya*, vol. 1, 211.

14. Sir I. Gardner Wilkinson, *A Handbook for Travelers to Egypt* (London, 1867), 151.

15. Ismail is also at least partially responsible for some of this development since he built al-Qubba for Tawfiq, and al-Zaafran palace in Abbasiyya for his mother. `Ali Mubarak, *al-Khitat al-tawfiqiyya*, vol. 1, 213; Mahmoud El-Gawhary, *Ex-Royal Palaces in Egypt* (Cairo: Dar al-Ma`arif, 1954), 85.

16. Nihal Tamraz, "Nineteenth Century Domestic Architecture: Abbasia as a Case Study" (M.A. thesis, American University in Cairo, 1993), vol. 4, 13–16,

17. Robert Ilbert, *Heliopolis, Le Caire, 1905–1922: Genése d'une Ville* (Paris: Centre nationale de la Recherche Scientifique, 1981), 116.

18. The other tram line utilized by foreigners was for the Pyramids. Broderick et al., eds., *A Handbook for Travellers*, 43.

19. *Al-Lata'if al-musawwara* (November 21, 1921): 4. Interestingly enough, fifteen years later the same magazine ran a similar cartoon, depicting the Minister of Transportation gruesomely squeezing a female Helwan in a press while a stylish Heliopolis looks on, see January 20, 1936, 6.

20. Evidence of this class comes not only from an expanded government that hired such bureaucrats, functionaries, judges, and teachers, but also from advertising, which demonstrates a growth in urban professions with numerous advertisements for private schools, teachers, tutors, governesses, doctors, dentists, lawyers, and urban services e.g., banking and insurance. See e.g., advertisements for St. Mary's Mission School (for boys and girls): *al-Muqattam* (October 1, 1896): 4; Bab al-Luq Girls' School, *al-Muqattam* (October 1, 1896): 1; Kamal School, *al-Liwa'* (November 30, 1902); Husayniyya School, *al-Liwa'* (January 10, 1903): 3; Mademoiselle Steiner Wilchelm Heino, *Anis al-jalis* 4, 11 (1901): 881; Shaykha Ismahan, *al-Liwa'* (February 8, 1904). Roughly, 5.4 percent of advertisements in a random sample of from *al-Mu'ayyad* between 1893 and 1914 were dedicated to urban services.

21. Here I am borrowing terminology utilized by Erika Rappaport to discuss the women's press and consumer culture in Britain. See her *Shopping for Pleasure: Women in the Making of London's West End* (Princeton: Princeton University Press, 2000), 111. Interestingly enough, this same terminology can be applied to *la-'Isha*, a popular women's magazine in Israel discussed in chapter 5 of this volume, despite the disdain for consumer culture in traditional Zionist ideology.

22. For a description of this process, see Abu-Lughod, Janet. *Cairo: 1001 Years of the City Vicorious* (Princeton: Princeton University Press, 1971), 132–139.

23. Abu-Lughod, *Cairo: 1001 Years of the City Vicorious*, 132–134.

24. Proximity to key locations and the notion of good neighbors were long-standing principles in Cairene history. As the old Arabic proverb goes, "Inquire about (search out) the neighbor before the house [*fattish `an al-jar qabla al-dar*]."

25. Muhammad al-Muwaylihi, *Hadith `isa ibn hisham aw fatra min zaman* (Cairo: Dar al-Qawmiyya al-Tafaa wa-l-Nashr, 1964), [fourth edition], 4–5. This novel originally appeared in serial format in *Misbah al-sharq* between 1898 and 1902, and Muwaylihi later revised and edited the collection of articles. Muwaylihi came from a family that was downwardly mobile in the nineteenth century, and thus

it is not surprising that he is a vehement critic of many of the changes that took place over the course of that century.

26. DW, MW, *nizarat al-ashghal* [NSH], box 1/6, Demande de Concession au Gouverment de S.A. Le Khédive de l'Enterprise générale de numerotage des maisons des villes du Caire et d'Alexandrie, May 29, 1883.

27. One British official remarked, e.g., that "[t]he streets in our suburbs are lighted on strictly economical principles." See Edward Cecil, *The Leisure of an Egyptian Official* (London: Hodder & Stoughton, 1921), 122.

28. Muwaylihi, *Hadith `isa ibn hisham*, 100.

29. De Leon, *The Khedive's Egypt*, 27.

30. "Winter Clothes and Taking Off the Overcoat," *al-Muqtataf* 15, 5 (1891): 338–339.

31. For information on the evolution of the aristocrat's household in Egypt, see Mary Ann Fay, "Women and Households: Gender, Power, and Culture in Eighteenth Century Egypt" (Ph.D. dissertation, Georgetown University, 1993), 166–171.

32. See e.g., "Decoration of the Home," *al-Muqtataf* 6, 6 (1881): 368; "Decorating the Table," and "Organization," *al-Muqtataf* 6, 12 (1882): 750–752; "Arranging the Salon," *al-Muqtataf* 13, 5 (1889): 329–331; "Libraries," *al-Muqtataf* 7, 8 (1883): 507–508; "A Library in Every House," *al-Muqtataf* 13, 9 (1889): 627–628; "Decoration of the Home," *al-Muqtataf* 16, 8 (1892): 564–565; "Home Furnishings," *al-Muqtataf* 18, 2 (1893): 121–122; Furniture of the Home and Its Arrangement," *al-Muqtataf* 21, 2 (1897): 137–138; "Proper Arrangement of the Home," *al-Muqtataf* 23, 3 (1889): 217–218; "The New House," *al-Muqtataf* 23, 7 (1899): 543–545; "The Wife's Desk," *al-Muqtataf* 27, 4 (1902): 385–386; "For Whom Is the Leadership of the Family," and "Administration of the Home," 32, 7 (1907): 571–573 and 576–577, resp.; "The Bathroom," and "The Bedroom," *al-Muqtataf* 32, 10 (1907): 844–846; "The Tasteful House and the Sitting Room," *al-Muqtataf* 55, 4 (1919): 430–431; "The Home of Tomorrow," *al-Hilal* 22, 6 (1914): 465. The recommended divisions of the home seem to replicate the change in private space that took place in the West prior to the eighteenth century as work moved outside the home and home life became more privatized. Fatma Müge Göçek, *East Encounters West: France and the Ottoman Empire in the Eighteenth Century* (New York: Oxford University Press, 1987), 24.

33. Qasim Amin, *al-Mar'a al-jadida* (Cairo, 1900); idem, *Tahrir al-mar'a* (Cairo, 1899).

34. Leila Ahmed, *Women and Gender in Islam* (New Haven: Yale University Press, 1992), 162. For a critique of how this image harmed women, see Lila Abu-Lughod, "The Marriage of Feminism and Islamism in Egypt: Selective Repudiation as a Dynamic of Post Colonial Cultural Politics" in her *Remaking Women: Feminism and Modernity in the Middle East* (Princeton: Princeton University Press, 1998), 243–269.

35. Both Fay and Marsot provide interesting details on the lives of elite Egyptian women and their role as buyers, sellers, and traders in property during the eighteenth century. Furthermore, they assert that these women had administrative responsibilities in the home. Fay states that a Mamluk woman "managed a household the size of a small hotel"; however, she provides no evidence for this claim. Instead, she draws an analogy between Mamluk women and aristocratic Western European women between the ninth and mid-eleventh centuries. Like their counterparts in Western Europe, she theorizes that they exercised power in the absence of a centralized state. Moreover, she argues that the logistics of the Mamluk system meant that men were frequently away from their homes, and thus one might then speculate that women were in charge. Fay, "Women and Households," 217–218, 295. Using anecdotal evidence, Marsot attributes the ability of elite women to establish charitable associations in the twentieth century to their experience of running households in the previous century. Afaf Marsot, "Revolutionary Gentlewomen in Egypt," in Lois Beck and Nikki Keddie, eds., *Women in the Muslim World* (Cambridge, MA: Harvard University Press, 1978), 265–266, 275. Edward Lane, on the other hand, reports that "in most families the husband alone attends to the household expenses," although he also points out that women are in charge of domestic affairs. See his *An Account of the Manners and Customs of the Modern Egyptians* (London, 1846), vol. 1, 259–260.

36. Perhaps we could speculate that for the upwardly mobile, the advice columns and women's magazines were even more important.

37. For a general history of women's education, see Zaynab Farid, *Ta`lim al-mar'a al-`arabiyya fi-l-turath wa-fi-l-mujtama`at al-`arabiyya al-mu`asira* (Cairo: Maktabat al-Anglo-al-Misriyya, 1980). See also, Sophie Babazogli, "l'Education de la jeune fille musalmane en Egypt" (Thesis, l'Ecole des Hautes Etudes Sociales, Paris, 1927) (Cairo: Paul Barbey, 1928). For a discussion of education in Mamluk Egypt, see Jonathan Berkey, "Women and Islamic Education in the Mamluk Period," in Nikki Keddie and Beth Baron, eds., *Women in Middle Eastern History: Shifting Boundaries in Sex and Gender* (New Haven: Yale University Press, 1991), 143–157.

38. Regarding the rise of compassionate marriage in Egypt, see Beth Baron, "The Making and Breaking of Marital Bonds in Modern Egypt," in Baron and Keddie, *Women in Middle Eastern History*, 275–291.

39. Multigenerational households, polygamy, and even to a certain degree concubinage did not immediately disappear, but rather the nuclear family became the ideal. See e.g., "The Family," *al-Sufur* (January 9, 1917): 2, in which the author praises the European family for its order, purity, and happiness, which [s]he attributes to the partnership created by the man and the woman in a nuclear family. The royal family tried to display an image of monogamy from the time of Tawfiq (r. 1879–1892), even where it did not exist. Ken Cuno, "Ambiguous Modernization: The Transition to Monogamy in the Khedival House of Egypt,"

paper presented at the Annual Meeting of the Middle East Studies Association, San Francisco, November 18, 2001.

40. Najib Hajj, "The Eastern Woman and the Western Woman," *Anis al-jalis* 4, 5 (1901): 658.

41. "Foreign Proverbs About Economy," *al-Muqtataf* 7, 8 (1883): 507; "Assessing the Expenses of the House," *al-Muqtataf* 14, 8 (1890): 557–558; "Spending Money," *al-Muqtataf* 14, 10 (1890): 703; "Economy in the Kitchen," *al-Muqtataf* 22, 11 (1898): 752–754; "Buying Necessities," *al-Muqtataf* 23, 5 (1899): 379; "The New Home," *al-Muqtataf* 23, 7 (1899): 543–545; "The Woman and the Expenses of the House," *al-Muqtataf* 26, 7 (1901): 689–691; "The First Principles of Household Organization," *al-Muqtataf* 32, 8 (1907): 668–669; "Economy of Expenses," *al-Muqtataf* 47, 6 (1915): 586.

42. "Administration of the House," *al-Muqtataf* 32, 7 (1907): 576–577; "Administration of the Mistress of the House," *al-Muqtataf* 23, 4 (1899): 296.

43. "Preferability of Electrical Light," *al-Muqtataf* 5, 1 (1880): 25; "The Gilded Electrical Bathtub Heater," *al-Muqtataf* 6, 3 (1881): 159–160; "Household Chemistry," *al-Muqtataf* 8, 3 (1883): 179–181; "The Leisure of the Mistress of the House," *al-Muqtataf* 14, 1 (1889): 57–58; "New Styles of Cooking," *al-Muqtataf* 18, 1 (1893): 38; "Electricity for Cooking," *al-Hilal* 14, 5 (1906): 312; "Heating Food," *al-Muqtataf* 33, 6 (1908): 296; "Household Organization and Electricity," *al-Hilal* 24, 1 (1915): 82; "Mechanical Devices Instead of Servants," *al-Hilal* 24, 8 (1916): 681–682; "The Wondrous House," *al-Hilal* 29, 10 (1922): 982–985.

44. According to Yacub Artin, between 1882 and 1907, house rents increased from a ratio of one to two and a half and domestic servants' wages from one to two point one. See the extract from his *Essai sur les causes de renchérissement de la vie materielle au Caire au courant du XIXe siécle* in "The Trend in Prices, 1800–1907," in Charles Issawi, ed., *The Economic History of the Middle East* (Chicago: University of Chicago Press, 1966): 450–451. After the financial crisis of 1907, and during the years of the war, articles in columns, e.g., *bab Tadbir al-manzil* in *al-Muqtataf*, urged women to limit household expenses.

45. See e.g., "The Building of a House and Its Happiness," *al-Muqtataf* 6, 6 (1881): 368–369; "The Distinction of the Woman," *Anis al-jalis* 5, 1 (1902): 915–919; "Paradise of the Woman," *Anis al-jalis* 5, 3 (1902): 978–982; Muhammad Mustafa `Ajizi, "The Nation is Reformed Only with the Reform of Families," *Anis al-jalis* 6, 9 (1903): 1558–1559; "Advice from a Mother to Her Daughter on Marital and Household Mattters," *al-Sa`ada* 1, 8 (1902): 169–170; Habib Ma`ushy, "The Woman's Dominion over the World," *Anis al-jalis* 7, 4 (1904): 1773–1783; Najib `Awwad, "The Life of the Married Couple," *al-Sa`ada* 2, 3 (1903): 476–479 and 3, 9/10 (1904): 593–596; "Knowledge and the Woman," *al-Sa`ada* 2, 9/10 (1904): 596–604; "The Man and the Woman," *Anis al-jalis* 9, 7 (1906): 165–166; "Characteristics of the Woman" *Anis al-jalis* 9, 9 (1906): 270–272; "For Whom Is the Leadership of the Family," *al-Muqtataf* 32, 7

(1907): 571–573; "Quranic Commentary [on the Equality of Men and Women]," *al-Manar* 12, 5 (1909): 331–332; An anonymous Egyptian (male) doctor, "Education of the Egyptian," *al-Mu'ayyad* (February 19, 1914): 2; "The New Woman in the East and the West," *al-Hilal* 30, 3 (1921): 222–231; "The Man's Love and the Woman's Love," *al-Hilal* 31, 5 (1923): 486–489; Muhammad Zaki `Abd al-Qadir, "Where Is Happiness?," *Ummahat al-mustaqbal* 1, 5 (1930): 161–165.

46. See e.g., "The Woman's Rights," *al-Muqtataf* 7, 1 (1882): 17–22; Maryum Nimr Makarius, "Raising Children," *al-Lala'if* 3, 3 (1888): 97–104; Jirjis Effendi Hanna, "The Status of the Mother," *al-Muqtataf* 14, 6 (1890): 404–408; Tawfiq `Aziz, "Is the Upbringing of Children Dependent on the Mother or the Father," *al-Hilal* 1, 5 (1893): 213–214; John and Niqula Haddad, "Is the Upbringing of Children More Dependent on the Mother or the Father," *al-Hilal* 1, 7 (1893): 322–324; "Household Organization," *Anis al-jalis* 2, 3 (1899): 110–112; An anonymous female writer, "Knowledge and the Woman," *Anis al-jalis* 2, 4 (1899): 142–145; Muhammad Mustafa `Ajizi, "The Nation is Reformed Only with the Reform of Families," *Anis al-jalis* 6, 9 (1903): 1558–1559; "Complaint of Mothers About Raising Daughters," *al-Muqtataf* 28, 10 (1903): 874–878; "Knowledge and Women," *al-Sa`ada* 2, 3 (1903): 480–483, intoned in 2, 7/8 (1904): 561–564 and 2, 9/10 (1904): 596–604; "The Woman's Education," *al-Sa`ada* 2, 7/8 (1904): 566–567; "The Woman, the Conscience, and Upbringing," *Anis al-jalis* 6, 3 (1903): 1355–1359; "The Wife's Beauty and the Mother's Beauty," *al-`A'ila* 3, 9 (July 4, 1904): 65–67; "The Shame of an Ignorant Mother," *al-Tarbiya* 1, 2 (1905): 13; "The Nation [*al-umma*] is a Fabric of Mothers, So We Must Educate Daughters," *al-Hilal* 15, 4 (1908): 139–143; "Duties of the Woman," *al-Jins al-lat'if* 1, 3 (1908): 87; T. Hanayn [female student], "The Ignorance of Mothers in Proper Education," *al-Jins al-lat'if* 1, 4 (1908): 110–114; "The Woman, Between Prostitution and the Veil," *al-Hilal* 19, 2 (1910): 106–109; Labiba Hashim, "Upbringing" [speech given at Egyptian University] reprinted in *al-Muqtataf* 38, 1 (1911): 274–276; "Who is More Worthy of Education First, Fathers and Mothers, or Boys and Girls," *al-Mu'ayyad* (February 1, 1914): 2. For a discussion of the educated mother in the Iranian context, see Afsaneh Najmabadi, "Crafting an Educated Housewife in Iran," in Lila Abu-Lughod, ed., *Remaking Women*, 91–125.

47. For information on the European context see Karen Offen, "Liberty, Equality, and Justice for Women: The Theory and Practice of Feminism in Nineteenth Century Europe," in Renate Bridental, Claudia Koontz, and Susan Stuard, eds., *Becoming Visible: Women in European History*, 2nd edition (Boston: Houghton Mifflin, 1987), 346. On the British occupation, see Leila Ahmed, *Women and Gender in Islam* (New Haven: Yale University Press, 1992), 150–153.

48. In 1911, there were 25,023 girls in schools under the inspection of the Ministry of Public Instruction, which included some *kuttabs*, higher primary schools,

a School of Practical Housewifery, and teachers' schools. Parliamentary Papers, *Report by His Majesty's Agent and Consul-General on the Finances, Administration, and Condition of Egypt & the Soudan*, 1911 (London: Harrison & Sons, 1912), 26. This figure does not represent female attendance at schools run by missionaries, Muslim charitable organizations, or the minority communities of Egypt. The American Mission alone had more than 4,000 female students in its schools in 1906. Charles Watson, *Egypt and the Christian Crusade* (Philadelphia: United Presbyterian Church of North American, 1907), 278. The rapid spread and visible appearance of the American Mission reinvigorated other groups, e.g., the Jesuits, the Scottish Mission, the Church Mission to the Jews, as well as the resident Muslims, Copts, Jews, Greeks, Italians, and Armenians. For statistics on schools and students, see James Heyworth-Dunne, *An Introduction to the History of Education in Modern Egypt* (London: Frank Cass, 1968) or Amin Sami, *al-Ta`lim fi-l-Misr fi sanatay 1914–1915* (Cairo: Matba`at al-Ma`arif, 1916).

49. Rifa`a Rafa`i al-Tahtawi, *Manahij al-albab al-Misriyya fi manahij al-adab al-`asriyya*, 2nd edition (Cairo: Matba`at Shirkat al-Ragha'ib, 1912), 350–351 and idem, *Murshid al-amin li-l-banat wa-l-banin* (Cairo: Matba`at al-Madaris al-Malakiyya, 1289 AH [c.1872]), 6, 32, 48–49, 66–68, 91, 101–106, 120–121, 134, 195–207, 215–256, 273–277, 372–373.

50. For more information on the issue of education under the British occupation see my "Competing, Overlapping, and Contradictory Agendas: Egyptian Education Under British Occupation, 1882–1922," *Comparative Studies of South Asia, Africa, and the Middle East*, forthcoming.

51. Since many textbook authors were Syrian Christians or Copts, it was far more common to speak of Eastern values, rather than Egyptian or Muslim values. Furthermore, Eastern was a uniting and distinguishing feature as opposed to the alien construct of Western.

52. Antun al-Gamayyil, *al-Fatat wa-l-bayt* (Cairo: Matba`at al-Ma`arif, 1916), 7.

53. Al-Gamayyil, *al-Fatat wa-l-bayt*, 33.

54. Francis Mikhama'il, *al-Nizam al-manzili* [*NM*] (Cairo: Matba`at al-Ma`arif, 1913), 66–85.

55. Mikha'il, *NM*, 32–51, 66–85; idem, *al-Tadbir al-manzili al-hadith*, part one [*TMH1*] (Cairo: Matba`at al-Ma`arif, 1910), 36.

56. Al-Gamayyil, *al-Fatat wa-l-bayt*, 21–23.

57. Francis Mikha'il, *al-Tadbir al-manzili al-hadith*, Part 2 [*TMH2*] (Cairo: Matba`at al-Ma`arif, 1910), 168.

58. Mikha'il, *TMH2*, 174.

59. Al-Gamayyil, *al-Fatat wa-l-bayt*, 23; Mikha'il, *TMH2*, 173–174.

60. Mikha'il, *TMH2*, 14–16.

61. I should also note that turn-of-the-century advertising in both the mainstream and women's press included such objects. See e.g., advertisement for Charalambo Diamondi [Asian imports], *Anis al-jalis* 3, 11 (1900): 438; advertisement for Widow Stanati's Jewelry and Antique Store, *Anis al-jalis* 3, 11 (1900): 437; advertisement for Imprimerie al-Tegariat, *al-Sa`ada* 2, 5 (1903), cover.

62. Mikha'il, *NM*, 12–14.

63. Mikha'il, *TMH1*, 185–196. Compare this to traditional visiting etiquette as described in Edward Lane, *An Account of the Customs and Manners of the Modern Egyptians*, 1: 277–281.

64. Mikha'il, *TMH1*, 26–28, 171–175.

65. Mikha'il, *TMH1*, 225–237.

66. Beth Baron, *The Women's Awakening in Egypt* (New Haven: Yale University Press, 1994), 67–71; Juan Cole, *Colonialism & Revolution in the Middle East* (Princeton: Princeton University Press, 1993), 128.

67. "The Benefits of Advertisements," *al-Hilal* 5, 11 (1897): 413–414; *al-Hilal* 4, 9 (1896): 344; "Advertisements in *al-Hilal*," *al-Hilal* 10, 5 (1901): 151.

68. *Anis al-jalis* 6, 12 (1903): 1637 and 1639.

69. For a complete discussion of this issue, see my "Creating *al-Sayyida al-Istihlakiyya*: Advertising in Turn-of-the-Century Egypt," *Arab Studies Journal*, 8, 2/9, 1 (Fall 2000, Spring 2001): 61–96.

70. See e.g., advertisement for Ahmad Kamil's store, *al-Sufur* 1, 7 (1915): 7; advertisement for Muhammad Amir's store, *al-Lata'if al-musawwara* (February 14, 1921): 10; Egyptian Clothing Company advertisements, *al-Lata'if al-musawwara*, ran throughout 1921–1922; advertisement for Ilyas `Addad's furniture store, *al-Lata'if al-musawwara* (October 16, 1922): 15.

71. Robert Tignor, "Bank Misr and Foreign Capitalism," *International Journal of Middle East Studies* 8, 2 (1977): 181.

72. The Cleopatra style figure was quite commonly used to represent Egypt in nationalist iconography. Palmolive soap advertisement, *al-Lata'if al-musawwara*, ran repeatedly throughout 1922.

73. Lactagol advertisement, *al-Lata'if al-musawwara*, ran throughout 1922.

74. For a discussion of the use of advertising a source for history, see Roland Marchand, *Advertising the American Dream: Making Way for Modernity* (Berkeley: University of California Press, 1985), xviii–xx.

75. The products included a hair dye, complexion creme, weight loss medication, and "Eastern" pills guaranteed to give a woman the bust of a goddess. All of these advertisements required response to an address in France. See *al-Muqattam* (April–May 1892). It should be noted that it was quite common for patent medicines to be available by mail. Thus, use of the post office box was not a means of gauging response, but rather was a common means of conducting business.

76. Lisa Pollard, "The Family Politics of Colonizing and Liberating Egypt," *Social Politics* 7, 1 (2000): 47–79.

77. Peter Gran, *Islamic Roots of Capitalism, 1769–1840* (Austin: University of Texas Press, 1979), 5–8, 18, 21; F. Hasselquist, *Voyages and Travels in the Levant in the Years 1749, 1750, 1751, 1752*, trans. (London, 1766) as cited in Roger Owen, *The Middle East in the World Economy, 1800–1914* (London: I. B. Tauris, 1993), 9. It should be noted however, that Egypt's commercial ties to Europe date back to the ancient world. See Janet Abu-Lughod, *Before European Hegemony: The World System A.D. 1250–1350* (New York: Oxford University

Press, 1989). What changed in the eighteenth century was demand for grain and other raw materials given new demographic exigencies and developments brought on by the beginnings of the industrial revolution.

78. For a discussion of how this process worked with respect to fashion, see Nancy Mickelwright, "London, Paris, Istanbul, and Cairo: Fashion and International Trade in the 19th Century," *New Perspectives on Turkey* 7 (1992), 134.

79. Ahmad `Izzat `Abd al-Karim, *Tarikh al-ta`lim fi misr fi `asr muhammad `ali* (Cairo: Matba`at al-Nahda al-Misriyya, 1938), 670; Church of Scotland, *Home and Foreign Missionary Society* (Edinburgh, 1859), Vol. XVI, 134 as cited by J.H. Sislian, "Missionary Work in Egypt During the 19th Century," in Brian Holmes, ed., *Education and the Mission Schools: Case Studies in the British Empire* (New York: Humanities Press, 1967), 198–199.

80. Watson, *Egypt and the Christian Crusade*, 223–224.

81. Finney, "Kindergarten Schools in Egypt, the Vision," in *The Child in the Midst: The Story of the Beginning and Development of the Kindergarten in Mission Schools and in the Government Schools of Egypt* (n.p./n.d.).

82. For detailed statistics on students and schools see Charles Watson, *Egypt and the Christian Crusade* and Andrew Watson, *The American Mission in Egypt, 1854–1896* (Pittsburgh, 1898). See also James Thayer Addison, *The Christian Approach to the Moslem* (New York: Columbia University Press, 1942), 143–144 and Grace Thompson Seton, *A Woman Tenderfoot in Egypt* (New York: Dodd, Mead & Company, 1900), 100. For a personal account of missionary education's impact, see the (auto)biographical sketch by Jean Said Makdisi entitled "Teta, Mother, and I," in Suad Joseph, ed., *Intimate Selving in Arab Families: Gender, Self, and Identity* (Syracuse: Syracuse University Press, 1999), 36, 47–50.

83. For detailed information on Ismail's buying habits, see DW, PI, dossiers 10/3–10/17.

84. According to one foreign visitor, some women of Ismail's harem even wore the Khedive's picture set in diamonds as a piece of jewelry on their shoulder. Mrs. William Grey, *Journal of a Visit to Egypt, Constantinople, the Crimea, &c. in the Suite of the Prince and Princess of Wales* (New York, 1870), 39.

85. Although there were some European tutors and governesses dating back to the reign of Muhammad Ali, the process was regularized under Ismail and spread more widely among the elite. Yacub Artin, *L'Instruction Publique en Egypte* (Paris, 1890), 129–130, 134; `Abd al-Karim, *Tarikh al-ta`lim fi misr fi `asr muhammad `ali*, 673–674.

86. With respect to the bureaucratic class, see Robert F. Hunter, *Egypt Under the Khedives, 1805–1879. From Household Government to Modern Bureaucracy* (Pittsburgh: The University of Pittsburgh Press, 1984), 100. With respect to wider adoption, it is apparent through advertising, women's columns and magazines, as well as textbooks.

87. André Aciman, *Out of Egypt* (New York: Riverhead Books, 1994), 132.

88. Hans Wehr, *A Dictionary of Modern Written Arabic*, 3rd edition (New York: Spoken Language Services, 1976), 72.

PART II

Inside-out, Outside-in,
Representations of Domestic Spheres

CHAPTER 3

Late Ottoman Photography: Family, Home, and New Identities

Nancy Micklewright

N ews of the art of photography reached Istanbul shortly after its invention in 1839, its presence announced in Istanbul in the newspaper *Takvim-i Vekayi* on October 28, 1839.[1] Within two decades, there were many commercial photography studios in the city, owned by Greeks, Armenians, and Europeans resident in Istanbul. Muslims were slower to learn the new technology, although with the incorporation of photography into the curriculum of the Imperial School of Engineers, Muslims too began taking photographs.

The Ottoman interaction with photography apparently demonstrates many similarities with its use by the European and British consumers who had a more direct link with the invention of the new medium. Since photography reached Istanbul so soon after its invention, Ottoman and European consumers were exploring the artistic, communicative and documentary possibilities of the photographic image at the same time in the development of the medium. However, in the absence of specific proof, it is impossible to assume similar, culturally nonspecific meanings for Ottoman and European photographic imagery, an important point to which I will return later.

Photographic images of the Ottomans produced for outsiders are much better known than images by the Ottomans of themselves. Found in both public and private contexts, photography was a means by which Ottoman men and women could control their representation of themselves and their society for a particular audience. Thus we find Ottomans of different social classes having their portraits made to be saved in family albums or distributed in *carte-de-visite* or cabinet card format to friends and associates. On a more public level, the Ottoman government assembled hundreds of albums of photographs on a huge variety of topics, for presentation to foreign governments or at the behest of the sultan, as a means of documenting aspects of the empire. These uses of photography are a rich source for understanding some of the ways in which the Ottomans sought to represent themselves as they confronted modernity.

As commodities, photographs occupy an interesting position, extremely cheap to produce and acquire, yet valued because of their long postproduction life and circulation. In the Ottoman context, the market for photographs was complex, responding to very diverse kinds of consumer demand. This article examines various aspects of photographic production in the Ottoman context, comparing the well-known photographs of the Ottomans that have appeared in publications of Middle Eastern photography to the lesser known corpus of images produced by (or for) the Ottomans themselves for a variety of uses. I am particularly interested in exploring the ways in which Ottomans experimented with photography as a means of defining new social identities. I assert that photography allowed a diverse group of Ottomans to take some measure of personal agency in creating their own representations. In addition I believe an interaction with photography may be a means of signaling modernity for Istanbul consumers from different ethnic and social classes. My approach to the images is based on close visual analysis, as well as a careful contextualization of photography in the context of economic production and consumption patterns. In the late Ottoman world, photography was a new and extremely influential means by which issues of personal and social identity were negotiated.

As I have noted elsewhere, photographs occupy an intriguing position in consumption studies, and indeed in the material world of economic activity.[2] Their relatively modest cost places them at the bottom of a scale of relative economic value, especially compared to such conspicuously expensive items as *yalıs* (summer homes along the Bosporus), elaborate dresses, furniture in the European mode, jewelry, and other similarly costly objects that are sometimes used as an index of consumer practices in an Ottoman context. In terms of cost, photographs are more similar to ephemeral goods such as food or drink, lower-priced goods, whose use nonetheless may reflect important changes in consumption practices. Yet far from being regarded as ephemeral, photographs are valued precisely because of their ability to capture a moment and preserve it. The modestly priced photograph, accessible to many levels of society, is saved and circulated long after *yalıs* have burned down or been sold, dresses worn out and old furniture discarded. In this way, the photograph is much more similar to other art products, part of whose value lies in the expectation of their long postproduction life and circulation.

Moreover, in the late Ottoman context, photographs were inextricably bound up with the discourses surrounding modernity. The possession and display of photographs and photograph albums was a means of signaling the desire to be modern, to have the modern home that is discussed for the Egyptian context by Mona Russell in chapter 2. The prescriptive literature discussed by both Russell and Lisa Pollard included information about photography, as did for example, the women's press in Istanbul. In the context of Pollard's conclusions about the transnational geography of modernity, it is interesting to note that Canadian women's magazines from the same period contain very similar advice about the significance of a clean, well-decorated home, as well as the responsibility of the mother of the family to document her family's history by having portraits taken of the children at regular intervals.[3] In terms of both technical dissemination (discussed here) and its role in constructing and presenting social identities, photography is a central component in the construction of modernity whose adoption globally does not follow the patterns that are generally assumed to have occurred in the typical construction of Orientalism.

Photography reached the Ottoman Empire within a few months of its invention, as daguerreotype artists traveled to Egypt and elsewhere to photograph antiquities, one of the first photographic projects envisioned by the French government when they announced Daguerre's invention in August, 1839. As suggested here, the Ottoman reading public was informed of the invention in a newspaper article a few months later. Commercial photographers arrived soon afterward to set up their studios in Pera, the neighborhood of the city frequented by Europeans. However, the presence of an 1845 newspaper advertisement in Ottoman placed by an Italian photographer, Carlo Naya, indicates that local residents as well as foreigners were perceived as potential clients by the foreign photographers.[4]

The tourist market was the initial impetus for the numbers of photographers who set themselves up in business in the major cities of the Ottoman Empire. In the days before tourists carried their own cameras,[5] they documented their travels by purchasing the work of commercial photographers, either as single images or in albums. Photographic studios were important stops on a visit to a new city, with the travel guides of the day providing information about the best shops for different kinds of photographs. By the end of the century, commercial directories from Istanbul were listing the names of sixty-five photographers active in the city, most with businesses in Pera.[6]

This brief description of the way in which the new art of photography reached Istanbul is not so different than the similar accounts that could be provided for many other places in the 1840s: New York, Montreal, Boston, Frankfurt, Cairo or Sidney. From the two centers of Paris and London, news of the new medium was communicated by letter, scientific journal, newspaper and the movement of people to cities all over the world. So although photographic technology was developed in Europe, its fast dissemination to far distant places allowed for the simultaneous exploration of the artistic, communicative and documentary possibilities of the photographic image in vastly different cultural settings.

Another factor that had an enormous impact on how photography was adopted around the globe has to do with the technologies involved in its invention. From our vantage point, more than 160 years after the beginning of photography, it seems easy to trace its technological history, following the development of certain processes that came to dominate the medium. But this is misleading. We have lived through similar patterns of technological development in the past few decades, for example in the initial introduction of two competing systems of home video technology, Beta and VHS. Eventually, despite significant advantages of the Beta system, VHS became the dominant technology. Looking back, the eventual dominance of VHS seems easy to predict, but at the time that the two competing systems were both in use, the outcome was not as obvious. In terms of photography, once the initial technology was disseminated, different, equally functional developments were made simultaneously by photographers working all over the world. A few eventually came to dominate the medium. It is important to remember that in the period under discussion here photographers were communicating among themselves about their work, publishing their processes and discoveries in photography journals that had a wide circulation. This ease of access to the technology sets photography apart from some other inventions of the eighteenth and nineteenth centuries that were disseminated from Europe and whose technology remained in the control of the original inventors.

A particularly sharp contrast to photography in terms of accessibility of technology is the sewing machine, another nineteenth-century invention that reached the Ottoman Empire from Europe.[7] I would argue that the camera and the sewing machine were similarly far-reaching inventions in the way they brought technology within reach of a wide range of population and introduced new modes of production to the region. However, despite the central position it came to hold in local and regional economies, control of the production and distribution of the sewing machine remained for many decades in the hands of its original inventors rather than local consumers. The same was not true of photography.

However, just as it is important to acknowledge the shared nature of the exploration of this new technology and art form, it is essential to contextualize the specific factors that distinguish the way in which photography was used in Istanbul from what went on elsewhere. So, for example, the circumstances surrounding the introduction and adoption of photography in Istanbul are quite different from those in more remote parts of the Ottoman Empire, other Ottoman cities such as Damascus or Jerusalem, or other parts of the Islamic world, such as Iran. In this chapter I am talking about what went on in Istanbul. This may turn out to be extremely similar to what happened in other large Ottoman cities, but it is premature to make that assumption. Istanbul, as the capitol of the empire and a very popular tourist destination, had a large population of resident and more transient Europeans, with a well-developed commercial network to serve their needs and to provide goods and services for the increasingly large number of Ottoman subjects who desired them. As a result, Istanbul residents had easy access to information about photography, to photographers, and to the technology itself.

As I noted earlier, the market for tourist souvenirs is credited with the growth of the photography business in the Middle East, and it is worthwhile to take a brief look at what kinds of images appealed to the visitors to Istanbul. Annie Lady Brassey, who traveled with her family to Istanbul in 1874 and four years later in 1878, is one such visitor, whose photographic collection is valuable in this context.[8] Lady Brassey's travels are documented in her travel writing (one volume describes her visits to Turkey and Cyprus) as well as a collection of seventy-three large photograph albums.[9] Volume sixty-two contains images of Turkey, a combination of professional and amateur work reflecting the Brassey's interest in what the city and its surroundings looked like, not who its inhabitants were or how they lived. Internal evidence suggests that many of these photographs were made by Lady Brassey herself. Her experiences in the city were all mediated through her identity as a British subject, closely identified with the political establishment. Her primary social contacts in Istanbul were members of the British diplomatic community. In their company she visited major architectural and cultural sites, the bazaars, and with other British women, called on a few high-ranking Ottoman women. Along with her husband, she entertained the Ottoman and foreign political elite on their yacht. The selection of photographs as well as their arrangement in the album signal the nature of Lady Brassey's contact with the Ottoman Empire. Following an initial survey of Istanbul, the balance of the album is arranged with one photograph per page, more or less following the same itinerary as the Brasseys did on their visits to Turkey. These photographs, mostly all architectural views or landscapes, include such images as an ambassador's *caique* just leaving the shore, a series of photographs of the newly completed Çırağan Palace, which Lady Brassey visited twice while she was in Istanbul in 1874, and numerous other views.

It is interesting to note that Lady Brassey's album, and indeed most of the other albums of the foreign visitors to the city which I have seen, rarely include many examples of the images of the harem women or dancing that are so often reproduced in modern studies of nineteenth-century photography in the Middle East. While these images certainly did find an audience, these views and others of Ottoman "types" were not so popular with the tourists whose albums I have seen. These visitors apparently preferred photographs of architecture and landscape, often devoid of people. This striking disparity among what we see represented in modern publications as typical nineteenth-century photographs of the Middle East versus what turns up in the albums of visitors to the region is a complex subject that I have addressed elsewhere and that is outside the bounds of this chapter.[10]

At the same time as the commercial photographers active in Istanbul were producing images for the tourist market, they were also working for the Ottomans themselves. Ottomans, in Istanbul and other major cities of the empire, were quick to understand the potential uses of the new medium, no doubt helped in large part by the examples set by the European visitors among them, and by their exposure to advertising. Individual Ottomans as well as the government explored the photographic image in a variety of contexts in the last decades of the nineteenth century.

The value of carefully assembled groups of photographs as a means for communication was clearly understood by some Ottomans. The Abdülhamid albums, assembled at the behest of the sultan for presentation to the American and British governments in 1893 and 1894, provide compelling evidence of the interest of the Ottoman government in using photography to convey a very particular view of the Ottoman Empire to targeted audiences.[11] The two sets of albums are nearly identical, each consisting of fifty-one volumes containing about 1,820 photographs. The photographs focus primarily on the most modern aspects of the empire: views of the modern city, important buildings, military installations, schools and their students, and accoutrements of the palace. So far we have no record of how these albums were received, it is obvious that the gift was initially conceived by Abdülhamid as a means of improving the negative image of the Ottoman Empire held by the American and British governments.

A second important photographic project carried out in an official Ottoman context is the Yildiz Albums, assembled for Abdülhamid, now in the Istanbul University Library. Containing approximately 34,879 photographs from all over the empire taken at his order, the albums are supposed to have allowed the sultan to keep abreast of events in his far-flung empire and abroad. While the albums contain the same range of views as the two sets that were presented to the British and American governments, they also include subjects that would not have been considered suitable for export, such as the new police stations that were built all over the empire and prisoners accused of various crimes. Despite significant overlap in their contents, the Yildiz albums were created for different reasons than those sent to the British and American governments, demonstrating the use of photography as a tool of official surveillance, which was emerging in other political contexts as it was being developed in the Ottoman Empire.[12]

To look now at what interested individual Ottoman consumers of photographs, we find that portraits of themselves and their families hold pride of place.[13] While they also produced and purchased other sorts of images, photographic portraiture of a kind very similar to what we might find being done by commercial studios in nineteenth-century London, Paris or Boston, found a ready audience among the Ottomans. The rise of photographic portraiture in the nineteenth century, whether in Paris or Istanbul, is a central aspect of the development of photography as a business entrenched in modern societies, accessible to nearly all classes of those societies. With the advent of photography, portraiture was no longer a one-of-a-kind mark of status, but an infinitely reproducible means of documenting one's social, familial or professional identity. All sorts of people, not only the elite, visited the commercial photographers' studios to have their portraits made. The photographic products that were wildly popular elsewhere, for example the card-sized *cartes-de-visite* that were collected into photo albums, as well as the photograph albums themselves are found in the Ottoman context.[14]

The album, as a vehicle for the viewing of portraits and other images, was a familiar construct in the Ottoman context. This had a role, I believe, in allowing the ready adoption of photographic portraiture by the Ottomans, since photography could be understood in the context of their own traditions of painted portraiture, long explored through the medium of miniature painting and displayed in albums.[15] The Ottoman response to photographic portraiture should be understood as an important example of a consumption pattern in which a foreign object, the photographic portrait, assumed a culturally specific meaning in the Ottoman context of portraiture, involving the substitution of a photographic portrait for the traditional painted miniature portrait in an album intended for circulation among a specific audience.[16]

The dominant aspect of traditional Ottoman portraiture, in which the image functions primarily as political message, is still present in Ottoman portraiture from the mid-nineteenth century onward. However other works testify to the existence of more personal images. These portraits involve the collaboration of the artist or photographer and the subject in the creation of a certain identity for the sitter, providing an opportunity for self-presentation not available earlier in Ottoman society. European style painting was adopted in Istanbul in the same decades that photography became available, and again, a more diverse group of people than would previously have been involved in portraiture, sat for painted portraits. In the Ottoman example, however, as opposed to what went on in Europe and North America, I would argue that photography was not introduced to respond to an already articulated desire for portraits, but rather led the way in expanding access to portraiture, and thus influenced the wider acceptance of the painted portrait.

Easy access to photographic studios and the inexpensive cost of a studio portrait opened the medium to those who could not have commissioned painted portraits. There are numerous surviving examples of studio portraits showing families nicely but not expensively dressed, having their portrait made, people whose story, even in this incomplete form would not necessarily have been included in any other medium. For example, in figure 3.1 we see a family facing the camera seriously, looking directly out at the viewer. Their clothes and shoes show that they are far from wealthy, and in fact virtually nothing is known about them other than what can be deduced on the basis of the photograph: a small, affectionate family (look at the way the two younger children are leaning on their mother), who have ventured into a commercial photographer's studio. Yet the existence of the photograph itself demonstrates a certain engagement with modernity on the part of this family and raises intriguing questions about the reasons they decided to have their pictures taken.

The new opportunities for self-presentation afforded by photography are particularly striking with regard to women's portraits, as the number of surviving examples of women's photographic portraits indicates.[17] We know that women frequently commissioned photographic portraits of themselves, or at least consented to have their portraits made. Numerous portraits of royal women have survived in the Topkapı Sarayı archives. However, we can also find depictions of Ottomans of other social classes, as for example in the

Figure 3.1 Turkish family in the photographer's studio. Photographer unknown. c. 1870. Private collection.

photographs of schoolgirls from the Abdülhamid albums or of the female relatives of middle-class families grouped together for a formal portrait in a photographer's studio.

Photographic portraits in particular, and perhaps to a lesser extent painted portraits, are part of the construction of a normative social reality that involves class, gender, ethnic and professional identities and claims to identities. A photograph is, or appears to be, tangible evidence for the existence of a specific person, whose age, economic status and social identity is revealed in the dress and pose chosen for the image. Yet the photograph, seemingly so permanent, is the work of a few minutes posing, wearing anything the subject may choose to put on. Trying on different identities can be as simple as trying on different clothes or choosing a different backdrop or setting for the photograph. In this period, photographers kept a selection of clothing, backdrops and props that were used in the creation of a particular image for their sitters. In the Ottoman context, when choices about dress and other aspects of self-presentation were important signals of modernity and social identity, the opportunity to experiment with different images, to try on different identities via the photograph was no doubt intriguing.

The easy variability of the photographic image has important consequences for how we read the photograph. While we all recognize that the possibilities for the manipulation of the photographic image are virtually unlimited, photographs continue to be valued, on an explicit or implicit level, for their truthfulness. Understanding the implications of the constructed nature of portraiture in the Ottoman context is challenging. In some cases, it is easy enough to detect the fabricated realities that photographs can create, as for example when European visitors to Istanbul don traditional Ottoman dress and pretend to be smoking water pipes, as in a well-known image by Abdullah Frères. Other images present more complex challenges. Wedding photographs often use the same formal characteristics of setting and posture as those produced in Europe, yet the relationship pictured in the photograph and the social circumstances of the subjects would have been dramatically different from their European counterparts. It is easy to be misled on the basis of the formal similarities of the photographs, the mode of representation, to assume corresponding similarities in content, in this case the social relationship pictured in the image. As these examples make clear for us, reading a photograph is a culturally specific practice.

In figures 3.2 and 3.3 we see another issue critical to the reading of Ottoman photographs. These two images each show an Ottoman woman dressed for the street, with *yaşmak* (the fine white veil and headdress Ottoman women wore at this time) and, in the case of figure 3.3, *ferace* (the long overcoat worn outdoors). The photographs resemble each other in their plain backdrops, three-quarter view of the subject, pose of the sitter with her face turned slightly away from the camera, and careful attention to details of dress. While each is a sensitive, careful portrait of the subject, the photograph in figure 3.3, by the Swedish photographer Guillaume Berggren, is being marketed as a type, yet another *dame turque*.[18] The woman in figure 3.2 is Refia hanım, whose portrait has been made by her husband, the Ottoman photographer, Ali Sami.[19] Distinguishing the commercial product from the private portrait is virtually impossible, based on the images alone. The awkward pose or obvious discomfort of the sitter that is often claimed as a means of identifying a tourist photograph is missing from Berggren's photograph. This pair of portraits illustrates an aspect of photography that has long frustrated those who seek to view photographs as unmediated images of reality. As should be clear by now, no photograph is unmediated. However, in the Ottoman case, when the same photographers and subjects were in shifting patron–client relationships with each other, reading a photographic portrait is particularly complex.

A second pair of photographs, both by Ali Sami (whose portrait of his wife, Refia hanım we examined earlier) provide an intriguing example of the way in which photographs were used as a means of displaying different identities. Figures 3.4 and 3.5 both depict the same young woman, Hamide hanım, daughter of the well-known Ottoman painter Hoca Ali Riza, and herself an amateur painter. In her portrait from about 1905, we see Hamide hanım as herself, dressed in the fashionable clothes worn by many Istanbul women at this time (figure 3.4). In the conventions of photographic portraiture common to the work of photographers in many parts of the world, she is pictured in a parlor setting, and with objects that are intended to allude to her interests, in this case a book, writing supplies and a paintbrush. The clock on the table is a traditional element from European portraiture that refers to the passing of time in the life of the subject. The small table, chair and lamp used to create the parlor setting are aspects of the new style of Ottoman domestic furnishings, which were gradually replacing the traditional Ottoman divans and low tables in this period. This photograph is a modern reprinting from the original negative. It is possible that when Ali Sami printed the image, he cropped it so that the wall and trees beyond the backdrop would not have been visible. But given his obvious interest in the use of photography to create particular settings, he may well have printed the full negative, as we see here, to emphasize the constructed nature of the image. In this photographic portrait, Hamide hanım and Ali Sami have together created an image of a serious, fashionable, artistic and educated young woman.

In figure 3.5 we see the same person, in a photograph from a few years earlier, facing the camera wearing a costume that evokes traditional Ottoman dress of several decades earlier, but is in fact not that, posed to create a parody of the popular commercial image of a harem woman found in contemporary Orientalist painting and photography. In this case, Hamide hanım strikes a suggestive pose, looking directly at the camera, with her hair down around her shoulders. She stands in front of a painted backdrop, which in this case fills the entire frame of the photograph. Hamide hanım and Ali Sami are playing here with an image of Ottoman women long familiar to visitors and residents of Istanbul, using the camera as their means of creating a fictional, temporary identity for Hamide hanım.

Figure 3.2 Refia hanım, wife of the photographer. Ali Sami. 1899. Private collection.

As Ali Sami and other Ottoman photographers took control of the camera, we see that they were completely at ease with the new medium, able to change the photographic identities of their subjects as easily as they could change their clothes. Far from remaining the passive subjects of tourist souvenir photos, Ottomans were adept at using the new medium to construct their changing identities. Indeed their use of photography can shift easily into the realm of social commentary as we see Hamide hanım and Ali Sami using comedy to disrupt a hackneyed stereotype.

The Ottoman engagement with photography (which for most Ottoman consumers equaled portraiture) was an important means for exploring new self-images and new identities, especially for women. Their appearance in photographic and painted portraits came at the same time as their new, more public voice in print, with more public roles for women poets and musicians, with increased educational opportunities for women, and the beginnings of the gradual entry of middle-class women into the workplace. In this period of intense social transformation, Ottoman women were not yet

Figure 3.3   Dame Turque. G. Berggren. Date unknown. Courtesy of the Moderna Museet, Stockholm.

actually visible in many contexts, but they were virtually present, to borrow a contemporary idiom, through their portraits. Thus women's portraits, and men's too, of course, were another means by which new identities, perhaps temporary ones in some cases, could be explored, for both the sitters and the viewers.

Ali Sami, whose work we have discussed earlier, is one of the best-known of the early Ottoman photographers. As photographic technology began to be more widespread and photographic supplies more generally available in Istanbul, the Ottoman themselves began taking pictures. Ali Sami, for example, was trained as an artillery officer, taught art and photography at the Imperial School of Engineers in Istanbul, and worked also as a military photographer under Abdülhamid. Indeed, a military education and employment by the court was initially the route by which Ottomans learned the new art.[20]

Once the camera was in their own hands, Ottoman photographers could take pictures of their family and home life, and thus have a greater role in

Figure 3.4   Hamide hanım as herself. Ali Sami. 1905. Private collection.

determining the manner in which they presented themselves photographically. Thus for example, in a 1908 image of Ali Sami's family (figure 3.6), they are depicted informally grouped in one of the rooms of their house, wearing everyday clothes, with four different newspapers, their titles visible, spread about the room. This family portrait, obviously a collaboration between photographer and family, offers a very particular view of the group as they chose to represent themselves. Similarly distinctive photographs by Ali Sami present a variety of carefully composed images, taken to record different moments in the lives of families or individuals, and reveal a considerable degree of sophistication in the extent to which some Ottomans were able to interact with the medium of photography.

An intriguing example of a family's self-presentation, Ali Sami's family portrait is of interest also for what it tells us about their domestic setting. We see them dressed in European style clothing, with one person seated in a chair, and

Figure 3.5   Hamide hanım dressed as a dancer. Ali Sami. c. 1900. Private collection.

with a marble-topped dresser in the background, on which are displayed two photographs. Another image of his family, not illustrated here, shows them in an interior room with furnishings in the traditional Ottoman style, that is, low divans that line the perimeter of the room. Just as the decades from 1850 to 1910 were a time of dramatic transformation in Ottoman dress and a variety of social habits, the ways in which the Ottoman home was organized also underwent significant changes.[21] This was a complex process, driven by a number of factors, among them the change in women's dress. As more Ottoman women began to wear the closely tailored, tightly fitting clothes of the European fashion tradition, their dresses (and more importantly, their corsets) did not allow them to recline easily, and they were no longer comfortable sitting on low couches with no back support. Chairs, unknown in traditional Ottoman interiors, became preferable.[22] As more European furniture entered the home, the

Figure 3.6 The photographer's family at home. Ali Sami. 1908. Private collection.

patterns of use changed. Traditionally, room furnishings consisted of the divan. Everything else was movable, to be brought out when needed and then put away. Bedding was stored in cupboards during the day, and meals were served on trays that were removed once the meal was over. As Ottoman houses began to be filled with dressers, beds, dining room tables and chairs, rooms were no longer multifunctional, but devoted to a single use. As Ali Sami's two photographs show, this transition in domestic interiors was a complex one, in which both traditions functioned simultaneously, at least to some degree.

With the arrival of European furniture in the Ottoman home, different habits of display were also adopted. In the traditional interior, there were few opportunities for arranging prized possessions for others to admire. These would have been seen in the context of their use: the beautifully embroidered napkins and elaborately decorated spoons and glasses brought out when tea was served, for example, often elicited comments from foreign visitors to Ottoman homes. Once dressers and side tables became a part of Ottoman interiors, the sorts of objects that filled those surfaces in English and American homes, also appeared in the Ottoman context.

To return to Ali Sami's photograph in figure 3.6, we see that among the objects displayed on the dresser in the back are two photographs, and indeed we know from other photographs and the numbers of surviving individual images and photograph albums that these were extremely popular among Ottoman consumers, at least in Istanbul. In Europe and North America, albums displaying family photographs or collections of *cartes-de-visite* were very common, claiming a place of honor on the parlor table in many middle-class households in the last decades of the nineteenth century. These albums, with their elaborate covers and pages intended to allow the display of *cartes-de-visite* and cabinet card photographs of celebrities, children and distant family members, were an important symbol of modernity.

Photographs and photograph albums, objects that found a ready acceptance among the Ottomans of Istanbul thus had multiple roles in the articulation of new identities in the late Ottoman world. As consumers of photographs, Ottomans visited the studios of commercial photographers to have their portraits made, thus taking the opportunity to display elements of their complex social identities (family, gender, class, ethnicity, religion, occupation), in the choices they made as they presented themselves to the photographer. In the form of *cartes-de-visite* and cabinet cards, they circulated their portraits, and their self-presentations, among their family and friends. As producers of photographs and albums, Ottoman photographers took an active role in articulating the ways in which social identities were defined and presented visually, in a variety of contexts. Finally, the display of photographs and albums became a part of the modern Ottoman home, a way of integrating the familiar (faces of friends and family) into the foreign, and at least initially unfamiliar, new domestic spaces.

Notes

1. Engin Çizgen, *Photography in the Ottoman Empire, 1839–1919* (Istanbul: Haşet Kitabevi, 1987), 20.
2. See Nancy Micklewright, "Personal, Public and Political (Re)Constructions: Photographs and Consumption," in *Consumption Studies and the History of the Ottoman Empire, 1500–1922*, ed. Donald Quataert (Albany: State University of New York Press, 2000), 261–288.
3. This material is discussed in Lara Wilson, *The Photo Album of Nellie L. McClung (1873–1951)* (M.A. thesis, University of Victoria, 1998).
4. See Çizgen, *Photography in the Ottoman Empire*, 64–66, for a reproduction of this advertisement and discussion of it.
5. The first relatively small, portable box camera was marketed by Kodak in the 1880s.
6. See William Allen, "Sixty-five Istanbul Photographers, 1887–1914," in *Shadow and Substance: Essays in the History of Photography*, ed. Kathleen Collins (Bloomfield Hills, MI: Amorphous Institute Press, 1990), 127–136.

7. As discussed in Uri Kupferschmidt's paper "The Social History of the Sewing Machine in the Middle East" for the workshop that was the precursor to this volume of essays.

8. Lady Brassey and her photograph albums are the subject of my forthcoming book, *A Victorian Traveler in the Middle East: The Travel Writing and Photograph Albums of Annie Lady Brassey* (London: Ashgate, 2003). In that book, I investigate the extent to which the individual circumstances and interests of British visitors to the Middle East determined the range of photographic images they collected during their visits to the region. Such an examination is crucial in understanding the extent to which nineteenth-century "Orientalist" stereotypes were in fact shifting and unstable. Lady Brassey's photographs of the Ottoman empire are presented here, briefly, as specific examples of a very large body of imagery (photographs produced for visitors to the nineteenth-century Middle East), which deserves more thoughtful analysis than it has generally received until now.

9. These are now in the collection of The Huntington Library in San Marino, California.

10. See Micklewright, *A Victorian Traveler in the Middle East.*

11. The Abdülhamid albums were the subject of a special issue of the *Journal of Turkish Studies*. See Carney E.S. Gavin, "Imperial Self-Portrait: The Ottoman Empire as Revealed in the Sultan Abdülhamid II's Photographic Albums," *Journal of Turkish Studies*, 12 (1988). They are also discussed in Selim Deringil, *The Well-Protected Domains, Ideology and the Legitimation of Power in the Ottoman Empire, 1876–1909* (London: I.B.Tauris, 1998).

12. See John Tagg's *The Burden of Representation: Essays on Photographies and Histories* (Amherst: University of Massachusetts Press, 1988) for groundbreaking work on the use of photography as a surveillance tool.

13. The determination of exactly what constitutes a portrait is a complex subject. For our purposes here we will consider it to be an image in which the representation of the sitter, a specific individual, is the main preoccupation of the artist.

14. Examples of *cartes-de-visite* and cabinet cards from the Ottoman context are reproduced in many works on photography in the Middle East. See particularly Çizgen, *Photography in the Ottoman Empire*; Gilbert Beaugé and Engin Çizgen, *Images d'Empire. Aux origines de la photographie en Turquie/Türkiye'de foto'grafın öncüleri* (Istanbul: Institut d'études françaises d'Istanbul, n.d.); and Nissan Perez, *Focus East. Early Photography in the Near East, 1839–1885* (New York: Abrams, 1988). Albums are reproduced less commonly, but see Beaugé and Çizgen for several examples.

15. There is an extensive literature on Ottoman painting, including portraiture. For the most recent work, and a complete bibliography, see Gülru Necipoğlu et al., *The Sultan's Portrait: Picturing the House of Osman* (Istanbul: Işbank, 2000).

16. Another important aspect of the Ottoman adoption of photographic portraiture is associated with developments in Ottoman painting more generally, the substitution of easel painting, done in oils or pastels for the older miniature painting

tradition. For a fuller discussion of this, and bibliographic references, see Nancy Micklewright, "Negotiating Between the Real and the Imagined: Portraiture in the Late Ottoman Empire," in İrvin Cemil Schick, ed., *M. Uğur Derman Armağani/M. Uğur Derman Festschrift* (Istanbul: Sabancı Universitesi, 2000), 417–438.

17. I have argued elsewhere that there are no true portraits of Ottoman women until the middle of the nineteenth century. Before then, in traditional miniature painting, women appeared in a limited number of contexts, and almost never as named individuals. For a fuller discussion of this subject, see Nancy Micklewright "Musicians and Dancing Girls: Images of Women in Ottoman Painting," in *Women in the Ottoman Empire: Middle Eastern Women in the Early Modern Era*, ed. Madeline C. Zilfi (Leiden: E.G. Brill, 1997), 153–168.

18. Berggren's career is the subject of the book, *Fotografiska Vyer Från Bosporen Och Konstantinopel*/[Photographic Views of the Bosphorus and Constantinople], by Leif Wigh (Stockholm: Fotografiska Museet, 1984).

19. See by Engin Çizgen, *Photographer Ali Sami, 1866–1936* (Istanbul: Haşet Kitabevi, 1989).

20. In the preface to her book *Photographer Ali Sami*, Çizgen recounts that her knowledge of the photographer and of his work is based on a chance phone call she received, which eventually led her to a collection of his glass negatives. While we know the names of some other early Ottoman photographers, a detailed understanding of their work awaits similar serendipitous discoveries of photographic collections and documents.

21. Obviously, these social transformations preceded 1850 and continued after 1910, but this was a particularly significant time of change for the aspects of consumer culture (dress, interior furnishings and photographs) discussed here.

22. For a more extended discussion of the change in women's dress and the relationship between dress and the adoption of European style furniture, see Nancy Micklewright, "Women's Dress in 19th Century Istanbul: Mirror of a Changing Society" (Ph.D. dissertation, University of Pennsylvania, 1986).

CHAPTER 4

Domesticity and Domestic Consumption as Social Responsibility in *la-'Isha*, an Israeli Women's Weekly, 1947–1959

Sonja Laden

T his chapter examines the role and standing of Israeli women's magazines in promoting an "Israeli" consumer consciousness and everyday life routines, in the years just prior to, and just after, Israeli statehood, 1947–1959, a period mostly associated with austerity, asceticism, economic hardship, and a scarcity of goods.[1] The underlying questions from which this chapter proceeds center on the evolution and sustained production of ostensibly marginalized, commercial print-commodities such as women's magazines, given their divergence from the officially promoted pre- and early-state Zionist–Socialist discourse and its hallmarks of native "Israeli identity."

Within this framework, my concern here is two-fold, namely to assess the viability of women's magazines as *products* of an "alternative" bourgeois culture within the modern Hebrew setting, and to inquire how they function as an *agency* that perpetuates, or promotes changes in, preferences, values, and specific models of conduct within the context of domestic life. I argue that Israeli women's magazines function as channels of cultural *importation* that transmit options typically believed to be excluded from, and seemingly incompatible with, official Zionist–Socialist discourse. In other words, they are viewed here as cultural artifacts in their own right, and as a mediating mechanism that plays a role in ordering everyday life routines.

Why Magazines Matter, and, How Domesticity Constitutes the Public Sphere

This chapter will demonstrate that aspects of an "alternative" bourgeois culture were operative even during the decade of official austerity in Israel, and that introducing a women's magazine into Israeli print-culture at the time provided a means of harnessing domestic space to the broader public sphere, where national and market concerns were at stake, without, however, fully relinquishing notions of personal selfhood. In light of this, the chapter is concerned with the evolution of everyday practices in the context of a "bourgeois" or "middle-class" Hebrew/Israeli culture, and examines the role of magazines in transmitting and establishing daily-life routines, including the consumption of the magazines themselves. Following Andrews and Talbot, Appadurai, Bourdieu, Carrier, Douglas and Isherwood, Du Gay and Pryke, Fine and Leopold, Jackson et al., Kopytoff, McCracken, Miller, Lury, Sahlins, Slater, Spigel, and Williams, this chapter presupposes that magazines, like all other artifacts consumed and used by a given community, not only "reflect" the sentiments and worldviews, nor simply "meet the demands" of the community, but in fact constitute a stand-alone factor in constructing it, through both the contents they transmit and their repeated use, which forms an organizing pattern.[2] As such, magazines are part of a shared repertoire available for, and imposed on, a community, shaping and constraining its possibilities for action.[3] Hence, they are important resources for tracing and assessing the specificities of everyday life practices and evaluating their net weight in ordering social life. As we shall see, they convey accounts of how the introduction of new commodities and consumer practices in Israel, including magazines themselves, actually came about and became operative in reorganizing Israeli society and culture, and provide important data about the concrete Israeli social spaces in which everyday transactions take place, where habits are acquired and performed.

Regarding the notion of domesticity, this chapter rejects the view often rooted in a simplistic, gendered separation of public and private spheres, in which the home is defined as a trivialized, feminine domestic realm, while the public realm is regarded as one of active citizenship, labor, and social usefulness.[4] Recently there has been a move away from this dichotomous notion of "separate spheres" in studies on the domestic consumption of commodities and commercial media.[5] Contemporary research suggests that an overall increase in the accessibility of commodities and services, especially in urban environments, attests to changing social norms with regard to domestic consumption (standards of cleanliness and comfort, heating in winter, hosting visitors as a regular social practice) and intra-household kinship, gender and familial relationships (the standardization of women working outside the home), as well as notions of individual well-being and ways of "coping with life" (e.g., by legitimizing a preoccupation with one's appearance through diet, beauty and hair-care routines, medical advice, and psychological counseling). This chapter extends this approach to women's magazines specifically: it aims to show how, regardless of whether women are housewives or work outside the home, magazines are instrumental in community formation, for they endorse the view of domestic practices and sentiments, including changing figurations of the family and households, as strongly intertwined with, and constitutive of, the public sphere. Even more so, following existing research on changing norms of domesticity in other cultures,[6] source material from Israeli magazines also suggests corresponding processes in Israeli culture.[7] Indeed, as the evidence in this article would

Figure 4.1   Recommended living-room layout (*la-'Isha*, May 21, 1947).

ארונות מטבח

מסבחינו הם על פי רוב קטנים ובלתי־ ארון לכלי מטבח. גם בדופן הדלת
נוחים. עלינו לנצל כל מטבח שבהם להח־ מקום לכלים.

Figure 4.2   Recommended kitchen layout (*la-'Isha*, June 25, 1947).

רהיט ישן — חדש יהיה !

Figure 4.3   How to restore old furniture (*la-'Isha*, September 3, 1947).

reveal, even before statehood, housing units in Israel were intended to provide more than just rudimentary shelter and little comfort (see figures 4.1, 4.2, 4.3, and 4.4 and analysis of such evidence given here).

Assuming that, like many other goods and consumer products, popular magazines in Israel, and their promoted contents, are adopted largely, though

מסיבות וארוחות חגיגיות

Figure 4.4   How to set the table for festive occasions (*la-'Isha*, May 13, 1947).

not exclusively, from Western cultures, their analysis allows us to trace processes of cultural change involving cultural contacts.[8] Hence, although the emergence of magazines in Israel occurred in circumstances different from the capitalist–industrialist development known to have motivated the evolution of magazines in the West, the cultural functionality of women's magazines in Israel is best understood against the historical backdrop of the Anglo-American magazine industry. While a full historical account of the impact of this industry on Israeli print-commodities is still pending, the cultural predominance of Anglo-American magazines is undeniable, and they can be assumed to have provided at least some kind of source for adaptation for Israeli and other cultures.

Late nineteenth-century mass magazines in America are typically held to mark the demise of the "genteel" reader, heralding a new stage in the commodification of American culture and the emergence of a new professional managerial class in America. A short while later, in the early years of the twentieth-century, they also introduced and institutionalized the notion of "muckraking," dedicated to exposing corrupt institutions and business practices in the name of "social democracy," seeking to convey a "growing awareness of the social character of human nature and a belief that public morality had to take precedence 'over private selfishness.'"[9] In England, domestic magazines and the penny-weeklies published for women in the 1890s are perceived to have heralded new notions of class (as suggested by the notion magazines for "women" rather than "ladies"), marking the emergence of working-class and socialist political activity, new constructions of middle-class consumption, femininity, and domesticity, new notions of the female body, the notion of a feminized press and new journalistic opportunities for women, and the new discursive mode of journalistic writing labeled the "new journalism," which stressed "human interest" stories, and was often represented by the "tid-bit" rather than the extended article, and displayed a subjective, intimately "personal" tone rather than a purely informative or "objective" voice.[10] Israeli women's magazines clearly bear resemblance to, and manifest continuities with, both English and American manifestations of the late-nineteenth century popular magazine (which in many ways over-lap), even as they are also part of the system of Hebrew language print-media published both before and after the establishment of the State.

Although compared with international scholarship research on women's magazines in Israel is relatively sparse,[11] existing studies do appear to share, however implicitly, the view that consumer magazines (i.e. those sus-tained by advertising revenues) may act as compelling means of inculcating understandings and images of womanhood by offering women concrete instructions for recommended social and individual conduct. The relative sparseness of inquiry into magazines in Israel may well be due to the fact that the field of magazine production in Israel is not yet considered a prominent factor in shaping market forces, nor is it seen as a self-regulated, autonomous source of cultural authority. Only recently have Israeli magazines begun to show signs of becoming a mass cultural form in which publishers and advertisers are able to traffic freely, as has been the case from at least the late nineteenth century on in America and England.[12]

Nonetheless, magazines were established in the cultural setting of the pre-state Jewish *Yishuv* in Israel, and included, for example, *ha-'Isha, Dvar ha-po`elet*, as well as *Kolno`a, and Tesh`a ba-`erev*.[13] There was an obvious dif-ference between the underlying mechanism of production of publications sponsored by Workers and Zionist organizations,[14] and privately owned commercial publications. Since the late 1940s the presence of magazines in Israeli culture has been on the rise, yet unlike typically canonized or overtly political print sources such as literary works and newspapers, popular maga-zines have yet to become a primary object of interest where the making of Hebrew, later "Israeli" culture and personhood are concerned. This is appar-ently due to the fact that magazines were viewed as part of an urban, "bour-geois," non-Israeli culture, typically marginalized in the historiography of the *Yishuv* and early statehood. Like other similar consumer-oriented artifacts and practices assumed to exemplify this non-Israeli culture, magazines have not been regarded as formative, nor as representative, of the prevailing ideological issues involved in the Israeli/Hebrew nation-building project.

In what follows I focus primarily on *la-'Isha*, the only women's weekly in Israel, also considered the most widely read Israeli periodical. Having made its debut on January 22, 1947, in the form of a supplement to the privately owned newspaper *Yedi`ot-akhronot* (founded in 1939), and launched separately on July 23, 1947, it is also the most enduring one. Given the regularized publication of *la-'Isha* over the past fifty five-years, it may rightly be viewed as one of the more "successful" print-culture commodities published in Israel.[15] The article tries to assess the extent to which the stance of *la-'Isha* deviated from the context within which it emerged, given the prevalence of national concerns before and after the establishment of the State, and the austerity and asceticism pertaining to the official national Israeli ethos. Evidence mentioned here suggests that introducing consumer practices and consumer consciousness into the domestic domain through *la-'Isha* by no means entailed excluding explicit manifestations of ideological and national concerns in the magazine. It should also be noted that this article is only able to provide a narrow, inside-view of readers who regard themselves, in one way or another, as "belonging" to mainstream "Israel Society," and therefore cannot adequately address the delicate complexities of minority group readerships.

A close examination has been conducted of a representative sample of fifty issues of *la-'Isha* dating from 1947 to 1959. Noteworthy columns and articles featured in this sample include guidelines and illustrations for the actual physical reorganization of rooms, a weekly menu of balanced main meals based on rations, a beauty column (instructing readers on skin care, use of make-up, and effective dieting), a child and babycare column, articles based on readers' personal experience and anecdotes, consumer reports and information, an illustrated fashion section modeled on American magazines, moralistic commentaries on prevailing social practices, profiles (photographic and textual) of successful women, a photography competition, a story format editorial with a didactic moral, a society-cum-gossip column, an advice column on intimate marital issues, health and personal well-being, and a growing range of advertisements. Some of *la-'Isha's* columns and features have persisted over the decades, while others have undergone changes or disappeared, or been replaced by new ones. Specifically, advertisements and advertising discourse have increased and changed radically over the years, even during the time-span covered by this chapter.

The Advent of a Public Discursive Forum for Women

From the outset, *la-'Isha* has provided a public discursive forum for Israeli women to comply with and critically posture themselves toward. As such, *la-'Isha* was also instrumental in consolidating a firm link between domestic and public spheres.[16]

La-'Isha's ostensible digression from written journalistic discourse in Israeli print-media at the time of its inception is best understood in terms of the rise of a quotidian, written journalistic discourse in Hebrew in the Israeli print-media. Although, as implied earlier, existing models of Anglo-American women's magazines were known to be increasingly operational from the late nineteenth century through to the 1940s,[17] and were presumably drawn upon in the Israeli context, a more germane local precursor of *la-'Isha* should be noted, namely the periodical *ha-'Olam ha-ze* [This World], a general-interest weekly founded in 1937 by Uri Keisari.[18] Despite its short-lived, relatively meager success, *ha-'Olam ha-ze* was seminal in consolidating a new journalistic model for periodicals in Israel, through which a new repertoire of cultural and linguistic options was introduced and legitimized.

In 1950 Keisari sold *ha-'Olam ha-ze* to army veterans Shlomo Avineri and Shalom Cohen. From its inception *ha-'Olam ha-ze* broke new ground by using a range of journalistic genres and discursive styles previously considered unsuitable for use in the print-media of pre-state Israel, most notably diversifying into critical, investigative journalism, popular society columns, and nude photography, popular, personalized forms of commonplace wisdom and sociocultural "chatter," women's interests,[19] alongside explicit ideological debates on defense and political issues. Linguistically, copy in *ha-'Olam ha-ze* departed from accepted stylistic patterns of literate, written Hebrew, seeking rather to conventionalize in writing an acceptable register of quotidian, everyday Hebrew as it was perceived to be spoken by "people in the street."[20] The historical significance of *ha-'Olam ha-ze* for *la-'Isha* also lies in the way it signaled the growing autonomy of the Hebrew language print-media in pre-state Israel at the time.

Partial corroboration of the way new discursive and social options were made manifest in and through journalistic channels is found in a recent interview with well-known journalist and media personality, Mira Avrekh, who has been writing for *la-'Isha* and its parent publication, *Yedi'ot akhronot*, since 1949. In a special anniversary album commemorating fifty years of *la-'Isha*'s publication,[21] Avrekh recounts how she began to produce fashion illustrations after receiving a parcel of women's magazines from a relative in America. She began her journalistic career working on a voluntary basis as a fashion illustrator with the morning newspaper, *Herut*, and early in 1949 she was persuaded by the editor of *Yedi`ot akhronot*, Noah Moses, to begin working for a salary with *la-'Isha*'s fashion column. There she produced illustrations and also wrote the captions that accompanied them, roughly translated from American magazines. When asked to expand her writing beyond these captions, she protested, claiming that she didn't know how to write. *La-'Isha*'s first editor, Aharon Shamir (1947–52) told her "Simply write the way you speak!" She did, and soon had her own advice column, *Mira Meya`etzet*, which became a regular feature in *la-'Isha*. Avrekh later left *la-'Isha* and established *Yedi`ot akhronot*'s society column, the mythological *Mira Avrekh mesaperet `al...* that became an institution in its own right and a model for Israeli gossip column in the print-media.

As a leading representative of journalistic circles in Israel, Avrekh, presumably voicing perceptions shared with her colleagues at the time, points out that the prevailing stylistic standard of journalism was very different in the early years of statehood. In her words, it comprised a moralistic combination of settler pioneerism and intellectual elitism, which, she contends, was also narrow-minded, conservative, and generally awkward. She adds that politics held pride of place, political debates were foregrounded, even in personal columns, and photographs were used only for "serious" news reports. "People," according to Avrekh, "were altogether different then"[22] and so, we might add, were the ways in which they were socially represented. Avrekh is in fact pointing to the fact that peoples' occupations and social roles at the time were mediated through a very different repertoire or set of representational images and cultural options than they are today. She points out, for instance, that "intellectuals were busy milking cows on kibbutzim while cab drivers whistled symphonies," and that "people claimed to have had no interest whatsoever in persons as individuals, in the distinctively personal, or in 'human interest' stories. And if they happened to read them, they would have been too embarrassed to say so."[23] It further follows from Avrekh's comments that public figures aspired to project respectability through conduct that was modestly humble and shied away from the limelight, and that then, far more so than today, overt claims to fame were definitively socially incorrect and decidedly improper. Avrekh's recollection of the declared stance of social norms at the time appears to overlook her own role and agency in changing these very norms, without, however, detracting from the validity of her claims.

Avrekh's most famous scoop, originally meant for *la-'Isha* but finally published in *Yedi`ot akhronot*, was an interview with Paula Ben-Gurion where she spoke at length about her husband, Israel's first prime minister,

David Ben-Gurion. When Avrekh called her editor, Aharon Shamir, to tell him she had an interview with Paula, he was extremely circumspect, promising to publish it only if it wasn't too *personal*. Paula's story was spiced with both personal trivia and political details, so that anecdotal chit-chat about what Ben-Gurion ate for breakfast, his daily exercise schedule, and the fact that he wore slippers, was intermingled with more "serious" accounts of the moves leading up to the Declaration of Independence.[24] Despite being peppered with personal trivia, the "feminized" story must have complied with journalistic standards at the time, since it was published by *Yedi`ot akhronot*, the more "serious" journalistic option of the two. Averkh's anecdotal account is not only journalistically illuminating, it is historically instructive in pointing out how the daily-life routines of Israel's first prime minister, David Ben-Gurion, came to provide, albeit unintentionally, a crucial link in personalizing the life of one of Israel's most prominent public figures. Although by now this is a common journalistic device, its historical significance in providing a model for the kind of magazine journalism *la-'Isha* and other Israeli print-media came to publish on a regular basis cannot be overlooked.

Zionist Austerity and "Bourgeois" Consumerism

Despite rationing and the retention of controls on goods during Israel's early years, *la-'Isha* also confirms that there were inconsistencies in the state of affairs housewives had to contend with. While according to *la-'Isha* frequenting shops and cafes, using public transport and the bank in early-state Israel were apparently routine and commonplace, these were by no means self-evident practices, nor was regulating modes of conduct within them a straightforward matter. As late as 1954 we note an advertisement in *la-'Isha* urging people not to keep their money under the mattress, but rather to put it the bank (October 7, 1954); an item published later that year warns readers to "Beware of shop assistants paid to make nuisances of themselves" by pestering customers into buying merchandise they don't need (December 1, 1954). An entire article, formulated as a personal anecdote, from an issue of *la-'Isha* dated April 2, 1947, deliberates whether a mother should insist on her young daughter giving up her seat for an elderly passenger. The writer herself assumes the role of social gatekeeper, mediating between the readers and the woman and child in question, advocating that children should never be seated on a bus if adults are standing. Evidence from *la-'Isha* hence upholds the notion that although new public spaces and economic institutes were opened for women, their use was often accompanied by social tension and disagreements, and entailed conflictual understandings of how to go about them.

An overall rise in commodities and services offered, especially in the Tel Aviv area, is attested in the weekly consumer column where new commodities and services were surveyed in *la-'Isha*. Among other things, this column specified Tel Aviv's first automatic laundry, located in Dizengoff street (August 3, 1949), reported the local manufacture of pressure-cookers by the "Garo" factory (August 17, 1949), supplies details on the local price of a refrigerator—200 Israel Pounds—as compared to 50 Israel Pounds in America, and calls for lifting tax restrictions on household electrical appliances as a means of undermining black market trading and appealing to the egalitarian purpose of rationing (August 24, 1949). Another issue of *la-'Isha* dated just one month later (September 25, 1947), records the public outcry of housewives accusing merchants and tradesmen (fishmongers, in particular) of cheating them by raising their prices on the eve of the Jewish New Year, while incriminating fellow customers willing to pay the higher prices. An article endorsing the organized employment of household help appeared in *la-'Isha* (February 10, 1954), indicating that even at the height of austerity domestic help was sought. Noting that most candidates for this kind of piecemeal work were Jewish women from Arab countries, the writer also states that they are entitled to fair pay (500 to 600 *prutot* per hour), insurance, and social rights. Other issues of *la-'Isha* feature reports and recommendations of available cosmetics and household furniture. In all this we note an increasing, later accelerated, rise in material objects both in the house (as attested by advertisements for furnishings, electrical appliances, heaters, detergents, soaps, perfume, cigarettes, books, etc.) and for public use, as well as the institutionalization of privately owned leisure amenities such as cafes and cinemas.

Cafés and visits to the movies indeed provided a backdrop for many of the articles and anecdotes reported in *la-'Isha*, suggesting that frequenting them had indeed become routine for many people, although cafés were most notably frequented by particular types of people, namely, journalists, politicians, writers, performing and other artists. At the same time, however, data from *la-'Isha* also suggests that alongside the accelerated rise in commodities and the standardization of bourgeois practices both within and outside the home, there were still periods of scarcity where certain goods were concerned. For instance, the equitable distribution of commodities did not always comply with the most basic needs of the female population: an urgent appeal for more sanitary napkins and cotton wool was published in the issue of *la-'Isha* dated May 23, 1951. This raises (but leaves unanswered) the question of how rationing was regulated and whose interests were prioritized in the process.

As already mentioned, not all the domestic consumption practices rendered explicit in *la-'Isha* were carried out at home: many of them took place in a variety of other small scale-social figurations, in extended family and friendship networks, informal civil society establishments, such as cafés and restaurants, public means of transportation, training frameworks for women, charity and social aid establishments, entertainment and recreation centers, health-care centers and clinics, various trade, market and commerce localities such as banks, grocery stores, dressmakers and tailors, dry goods and household ware suppliers, booksellers, pharmacies, and later in a broad array of stores, supermarkets, and shopping malls, and workplace locales, like businesses, offices, or manufacturing environments. Although these practices were carried out in the public domain, they are telling with regard to personal, individual conduct and "ways of being" in the world, and the ways they enabled people to position themselves in relation to local, "domestic" practices, and vis-à-vis imported codes of conduct.

An illuminating example, from an issue of *la-'Isha* dated August 24, 1954, exemplifies how conduct within the public sphere is brought to bear on individual postures and social positioning, and in turn feeds back into the public sphere. This account of a conversation between several women at a café touches directly on the way a sense of crude, unrefined public behavior as definitively "Israeli" is formulated against individual experiences of alternative, imported codes of conduct. Recounting, as though verbatim, several incidents undergone by Michal, an Israeli returning after a long stay in England, the author here wishes to suggest that Israelis should be attentive to other, non-Israeli modes of social behavior, as Israeli social norms are by no means standard. Michal is protesting against the lack of respect and civility, and the overall discourtesy, of public behavior in everyday life patterns of consumption. She indicts (a) the rude, offensive behavior of an airport official; (b) the disrespectful bank teller who mistakenly gave her too much money, and when corrected, snatched the money away, and enraged with Michal, gave her the correct amount; (c) an argument between herself and the waiter at the café who believes that she is obliged to provide him with the correct change, not vice versa. Noting that "nothing like this could ever happen abroad," Michal has clearly internalized imported norms of Western consumerism whereby customers are treated with respect, as though they "come first" and "are always right." But by publishing this account, both its author and the editor of *la-'Isha* are suggesting that there is good reason to institutionalize these norms in the Israeli public sphere.

Interesting here is the role played by *la-'Isha* in institutionalizing the camera as a household appliance recommended for routine domestic use by Israeli women within the family circle. A means of exploring new ways of publicly "presenting the self," the camera exemplified how images of domesticity were able to enter and become standardized within the public sphere. Soon after the weekly's inception in 1947, a competition was launched encouraging readers to send in photographic portraits of their children, and the winning photograph was published each week. An article entitled "Let's Photograph Our Children" (December 21, 1949) features a survey by the writer into the frequency with which her friends took pictures of their children, the prices of "user-friendly" cameras and the aptness of possessing one in times of economic and other hardship. The article centers on a detailed description of a mother's joy and elation at being able to share memories of her son's childhood with the writer of the article, who strongly endorses this sense of the mother's delight and pleasure. While in keeping with imported, bourgeois perceptions of the modern woman's delight in her responsibility toward keeping a record of her family history through photography, the cultural agency of the photographic medium can also be directly related to the magazine's view of itself as an emergent visual medium whose subject matter is also, and increasingly, presented visually within the public sphere. In this sense, *la-'Isha* no doubt also sought to boost its (primarily female) readership's need for consuming photography and visual imagery, with a view to increasing their gratification from the magazine as a medium in its own right.

The evidence mentioned here supports the view that despite the explicit socialist ideology of the Israeli state-in-the-making, and the link maintained between state-building and economic development, nationalist concerns were by no means entirely opposed to the social principle of market exchange and private enterprise.[25] Although market concerns may well have been marginalized in official Israeli state discourse, and Israeli society may even be regarded as a "socialized" market based more or less directly on principles of "managed" capitalism,[26] the growing viability of Israeli society as a modern market society cannot be denied. In fact, the circulation of commodities and consumer practices has since been legitimized as part and parcel of the Zionist–Socialist agenda that included a deliberate and concerted effort to promote the provision and consumption of Israeli produced goods (*Totzert ha-'aretz*),[27] also evidenced in the Workers Organization publication for women, *Dvar ha-po`elet*.[28] This agenda, whereby choosing locally produced goods has been favored alongside procedures of cultural importation, has encouraged the growing centrality of consumer practices and commodities gradually introduced in Israel.

Domesticity as Social Responsibility

The last section of this chapter illustrates how *la-'Isha* functioned as a mode of cultural mediation that provided models of conduct for everyday life routines and domestic behavior. As mentioned earlier, the magazine was launched in January 1947 as weekly supplement to the daily *Yedi`ot akhronot*. Bearing the subtitle *Popular Weekly for the Home and Family*, it referred to the social roles and standing of Jewish women in pre-statehood Israel in terms of two underlying, and frequently interconnected, types or models of domesticity: (a) domesticity as a national concern, and (b) domesticity as a market concern. In both cases the notions of home and family, and women's roles within these, were mediated through a more or less fixed range of *everyday life routines*. At the level of content, the *national model of domesticity* was overtly enlisted in the editorial excerpt below. The *market or consumer model of domesticity* appears to be somewhat more tenuous. However, the very appearance of *la-'Isha* as a separate supplement, later weekly, targeting women readers was driven by an appeal to *domesticity as a market concern*, prioritizing consumption within and around the domestic environment, that is, the home and family, as a more or less autonomous factor in regulating new social practices.

In what follows the article presents the editorial production stance of *la-'Isha* with its inauguration (penned by Aharon Shamir, the weekly's first editor 1947–1952), in order to clarify how during its early years *la-'Isha* construed efficient buying as a form of social responsibility, a yardstick, in fact, of "good" citizenship.[29] Heralding the new publication, the editorial deployed rhetorical strategies intended to motivate a new readership of women by underscoring the importance of social cohesion and solidarity, and the contribution of women readers to the broader national consciousness crystallizing among the Jewish population in Israel at the time. This national model of domesticity corresponds with, and may well have been modeled on, a similar wartime mobilization of the women's press in British and American women's magazines and television in the 1940s and 1950s, whereby Anglo-American media sought to converge "the national interest" with a postwar "domestication" rhetoric.[30] The copy, which appeared on the front page of the supplement issued on January 22, 1947, reads as follows:

We are pleased to introduce *Yedi`ot akhronot la-'Isha* (Latest News for Women). Indeed, women feature in all facets of our public, social, and cultural lives. A woman may well share her husband's special interests in economics, sports, chess, literature, art, etc., and read about these in our newspaper. But a woman is deprived when it comes to the issues that comprise her own special world: managing her own home, educating her children, sewing her own dresses [translation my own].

The title *Yedi`ot akhronot la-'Isha* clearly targets those women readers who are already likely to have access to the established parent publication, itself directed at both male and female readers. The editorial appeals to its intended readership to provide a crucial link between the home and family, and the emerging national entity, without renouncing their own particular needs in the process. However, the rhetorical construction of this appeal is somewhat more sophisticated, deploying a series of correspondences aimed at maximizing the rhetorical effect: the editor points out that the new supplement for women is both a product and an extension of its parent publication *Yedi`ot akhronot*, just as women themselves are constituents of coupled entities and extensions thereof, and their emergent domestic sphere is both part of, and an expansion of, the public sphere. Implicit here is the understanding that the new publication, its intended women readership, and the evolving domestic sphere to which both belong are equivalent factors in the same equation, and are all harnessed to the overall "national interest." This national model is conjured up again toward the end of the editorial (see later), when the association between "Hebrew" women, their "homeland," and the hardships entailed in securing the latter are rearticulated. The text suggests, then, that "matters specific to the world of women" are not detached from, but rather extend the boundaries of the public sphere. As noted earlier, this strategy is an attempt, conscious or not, to evoke a sense of harmony, rather than disjuncture, between the supplement and its parent publication, between women and men, the family and the national entity, and between domestic space and the public sphere. In so doing the editorial stance aims to forge among the supplement's readers a strong sense of shared national aims and concerns, and promotes its own cultural functionality at the same time.

On another level entirely, yet in similar vein, we also find the craft of knitting, a domestic pastime par excellence, recruited to national ends. During the 1940s knitting became an emblem of the way Anglo-American (especially British) domesticity was called upon to recruit women into the national wartime effort,[31] and in similar vein, under British Mandate (pre-state Israel), *la-'Isha* too heralded knitting a national pastime. In a lengthy column that appeared in another issue of the supplement, knitting was overtly proclaimed a "national calling": women were urged to knit for themselves, for their families, and for soldiers in combat (February 5, 1947).

A telling excerpt from the column entitled "These Women" (*ha-Nashim halalu*), a regularly featured society-cum-gossip column in early issues of *la-'Isha*, states:

> There is something which unites us all as one, which characterizes all women in *Eretz Israel*, [an "insider" term designating pre-state Israel] regardless of class or sector— we all knit...in town and country, in rural settlements (in the *Moshava, Moshav*, and Kibbutz), at home and in cafés, at the movies and during theater intermissions. We host knitting parties, and sit enthralled in the pleasures of knitting. Let us remind all our knitting sisters—the craft is a hallowed one, symbolizing how we can best help others (February 5, 1947).

The stance toward the craft of knitting as both religious ("hallowed") and sensuous ("enthralled in the pleasures of knitting"), and as unifying in both literal (geographical) and metaphorical (national) senses, confirms and exemplifies how domestic concerns were indeed marshaled to national ends.

In the Anglo-American world, magazines have long since overtly construed women as primary purchasing agents where the family is concerned, and women's magazines today are often chastised for overtly soliciting women as consumers, largely through advertising. This is substantiated in the editorial passage cited earlier, which claims to be concerned with granting women access to their own "specific" interests, even as it provides a list of traditionally feminine tasks through which women in pre-state Israel could legitimize and enhance the domestic sphere. In other words, although in terms of journalistic content *la-'Isha* intended to diverge sharply from political journalism at the time, it aspired to legitimize the concern with women's roles in the domestic sphere as socially acceptable and responsible, through the cultural importation of an existing model known to feature centrally in postwar Anglo-American magazines during the 1940s and 1950s.[32] In response to the centrality of warfare and military conflict within the Israeli national project, *la-'Isha*, like other British and American magazines of the same period, sought to enlist women as both purchasers (consumers) and readers primarily by appealing to their sense of frugality, construing efficient buying, caring for well-nurtured and well-raised children, and cultivating a pleasing, fashionable appearance as forms of national responsibility and "good" citizenship, while promoting the supplement itself as a didactic tool in this regard:

> ...we will teach women in *Eretz Israel* how to cut expenses according to their own needs, how to be frugal yet still fashionable, how to feed their families sensibly yet inexpensively, how to find the blessed middle-mean between fussing over their children and neglecting them, etc. Young Hebrew women

and girls will also find matters of interest in *Yedi`ot akhronot la-'Isha*: how to freshen up last year's frock for the coming season for next to nothing, how to care for and enhance their looks, etc. In addition, *Yedi`ot akhronot la-'Isha* will try to identify the tasks and problems facing Hebrew women in their homeland (*Yedi`ot akhronot la-'Isha*, January 22, 1947; translation my own).

In promising to teach *Eretz Israel* women how to "make the most of their given circumstances" by trying to rise above economic hardship and austerity, feed their families well ("sensibly yet thriftily"), even as they continue to dress attractively, and raise their children responsibly (without indulging or neglecting them), this passage implies that home management and the proper raising of children are socially correct enterprises that promise to be rewarding in their own right, rather than commodified practices that are also signs of conspicuous consumption in their own right. Interestingly, sewing or dressmaking, although forms of home-production in themselves, are addressed even at this early stage as consumption-oriented pastimes attesting more to the style, rather than the fact, of caring for one's looks, and updating one's clothes by trimming or altering them (addressed notably to younger unmarried women). In other words, they are suggested as markers of style rather than necessity per se.

Although the stereotypical feminine "interests" that appear in the editorial are modeled on representations of women in Anglo-American women's magazines, more specifically "local" affairs and "Israeli" images of women are also discerned in *la-'Isha*. In later issues we note, for instance, an in-depth feature on an Israeli high school teachers' strike and its impact on female students (September 20, 1959), and in a later Jewish New Year's issue (September 27, 1959), a feature on the Top Ten Women of the Year list chosen by readers. The latter includes, not surprisingly, a brief profile of the reigning Israeli Foreign Minister, Golda Meir, as well as profiles of two "ordinary" Israeli women promoted as role-models, Sarah-Bella Buchman, named "citizen of the year," a successful religious business woman active in assisting less fortunate young religious couples in setting up home, and "mother of the year" Shoshana Mizrachi, a forty-two-year-old mother of sixteen whose contribution to demographic growth in Israel is both noted and saluted. The editorial stance here seeks to render traditionally "domestic" practices such as charity work and childbearing as socially responsible acts of citizenship. Once again, the boundaries between the domestic and the public spheres are rendered malleable.

In a variety of ways, then, editorial copy from *la-'Isha* seems to rearticulate that "matters specific to the world of women" are not detached from, but rather transcend, and thereby extend, the domestic domain into the public sphere, and vice versa. Noteworthy too in the earlier 1947 text is the way the editorial voice addresses readers in the third person, unlike the conventional mode of address in Anglo-American women's magazines of the same period, which is personal and intimate, a tone implied by addressing the reader in the second person. Whether this is because the editor of *la-'Isha* was male, and as yet unfamiliar with conventional modes of editorial address in Anglo-American women's magazines, or whether this was merely an oversight, is not clear at this point in time.

While the editorial excerpt from the 1947 issue manifests a relatively "soft" conceptualization of women as consumers, a later editorial (September 27, 1959) represents a changing social map with regard to the heightened acknowledgement of *la-'Isha* as a print commodity in its own right, commenting explicitly on a marked growth in advertising volume, and the reciprocal relations between the two. Calling attention to changes in the magazine itself, this same editorial informs readers about the increased volume of the present issue (from twenty-six to forty pages). Moreover, by asking readers "Do you find yourself paging through this issue from back-to-front?" the editorial stance implies a reader more skilled in the protocols of magazine-reading than previously was assumed, and addresses the reader as party to the marked increase in advertising space, and to its gratifying incorporation into the actual space of the magazine itself alongside other copy.

The perception of the domestic sphere as a social space tied to the public sphere is further illustrated in *la-'Isha*'s introduction, during the late 1950s, of separate pull-out supplements entitled *Kedai la-da`at* (Worth Knowing), frequently edited by Tekhiya Bat-Oren. A veteran journalist and editor with *la-'Isha* for many years, it is noteworthy that Bat-Oren has refused to cooperate with the present author since she "is now a feminist and believes that *la-'Isha* was ultimately detrimental to women in Israel" (personal communication, May 25, 2000). Given Bat-Oren's own perception of her supplements as contributing to stereotypical, conservative images of women, and precisely because they continued to promote these images, I believe these supplements served a different purpose at the time: they were motivated largely by market concerns within the domain of the Israeli print-media. Their publication, in other words, attests less to the continued imposition of feminine cultural stereotypes through *la-'Isha* as an *agency*, and rather marks a different moment in the history of *la-'Isha* as a *product* in its own right.

Attesting to *la-'Isha*'s institutionalization as an Israeli print-commodity with a "past" of its own, these supplements mark an editorial attempt to reenergize *la-'Isha* through a slightly different textual format, intended to ensure the continued loyalty of existing readers and attract new ones. This also suggests *la-'Isha*'s strengthened position in the Israeli market of print-commodities. This is further corroborated by the way this scrapbook-type supplement confers upon itself the status of a "book" (*sefer*),[33] rather than a periodical. At the level of content, these supplements, like magazines in general and *la-'Isha* in particular, are effective agencies of repertoire transmission through repetition, whereby the repeated circulation and production takes place of similar yet different versions of a more or less formulaic model.[34] Like the weekly itself, they supply readers with concentrated information and practical advice on the routine practices of home management, tending a family/household, food preparation and nutrition (highly significant in view of objective conditions economic hardship and austerity in Israel at the time), personal care and beauty routines, fashion and the art of appropriate dressing, knitting and other home crafts, manners, social graces, and codes of etiquette in the private and public spheres. Although many of Bat-Oren's recommendations are dated and "old-fashioned" by contemporary standards, they do illustrate how conducive the magazine form is to propagating repetition, in its very modes of consumption.

Let us observe first some of Bat-Oren's recommendations on manners, social graces, and codes of etiquette in the public sphere, many of which are clearly cultural imports from Western culture. Explicit instructions are provided for socially correct conduct at the cinema, the concert hall, or the theater (don't arrive late, be considerate of patrons already seated as you make your way to your seat, stay seated during the performance, don't converse loudly with other patrons, etc.), at restaurants (ladies should follow the waiter to the table, smoking is allowed between courses and the use of ashtrays is mandatory, ladies should be helped with their coats, etc.), visiting a hospital (only during visiting hours), comforting mourners, general advice with regard to public demeanor in a queue or public transport (give your seat to persons older than you or disabled), socially correct use of the telephone (don't call anyone at meal times, between two and four p.m. [Israeli siesta time], or late at night).[35] Other instructions relate to respecting privacy (always knock before entering, even when you know you're expected), being considerate of others (e.g., helping with parcels), refraining from shouting and using vulgar language (illustrated by the Hebrew equivalent of "sure!" [*betakh*]), show respect for other people's belongings, and more.[36] Other examples mentioned recommend practices confirming women's authority at home, emphasize that the domestic sphere constitutes a feminized space that is nonetheless part of a broader public domain. Among the social graces addressed is the art of conversation, in which context Bat-Oren informs readers that "It is generally the woman who sets the behavioral standards for other members of the family."[37] Urging the reader to conduct herself as "a demure, lady-like creature who never raises her voice, and is careful never to argue with her husband in front of the children" confirms that the underlying model here is imported, and most probably inspired by a Victorian-based English repertoire. For anyone familiar with Israeli behavioral norms, it will be apparent that there are vast discrepancies between this advice and accepted norms of conversational conduct in Israeli society, where it is often standard behavior to raise one's voice in conversation, regardless of the circumstances at hand. Moreover, given the contemporary climate in which there is an increased assertiveness of feminist sensibilities in Israel and elsewhere in the world, a recommendation such as this one is quite likely to be held in disdain today.

No less conspicuous is the importation of instructions for the "art" of laying the table (including explicit measurements for, and illustrations of, the required fall of the tablecloth from the table-top), and the etiquette of seating guests or oneself at one's own dinner table at home (instructing the reader to "stand to the left of her chair, and to wait for a male fellow-guest, or her host, to help her onto her seat").[38] In addition we note, "all dishes should be served to the left of the person dining, while coffee, water, and other beverages will be served from the right."[39] Although these recommendations are specifically intended to be implemented at home, they are conducive, like many others mentioned earlier, to constituting the home as a social arena, and as such warrant consideration as contributing to the broader public sphere.

Conclusion

In this chapter, I have tried to describe the role and standing of Israeli women's magazines in promoting an "Israeli" consumer consciousness and everyday life routines, in the years just prior to, and just after, Israeli statehood. Evidence from *la-'Isha* has been considered in terms of (a) the rise of *a quotidian, written journalistic discourse in Hebrew in the Israeli print-media*, which was crucial in facilitating the introduction into Israeli society of a repertoire of cultural and linguistic options *oriented toward personal and market driven concerns, rather than strictly state-centered, Zionist considerations*, and (b) the *creation of domestic space* and the *regulation of domestic consumption* within the broader context of an emerging national entity and new ways of life during the years of early Israeli statehood. I hope the evidence considered here goes some way toward illustrating that just as domestic space is both firmly linked to, and constitutive of, the public sphere in Israel, so the state's socialist–nationalist concerns were by no means entirely opposed to the growing viability of an Israeli modern market society.[40] *La-'Isha*'s significance hence lies in the way it has paved the way for and legitimized a *repertoire of domesticity* within the public sphere in Israel, and provided a discursive forum for Israeli women to take it up and position themselves toward. By mediating a range of shared, popular understandings of what domesticity in Israel entails at different points in time, *la-'Isha* sheds light on the way the social construction of Israeli identity cannot straightforwardly be reduced to strictly political acts, ideologies, and institutions.

Notes

1. In late April 1949, the Minister of Supplies and Finance Dov Yosef imposed an official policy of rationing. This period of government sanctioned austerity continued until February 1959, when last government restrictions on staple foods and supplies were withdrawn. See Mordechai Naor, *Book of the Century* (Tel Aviv: Am Oved, 1996) (Hebrew).
2. Maggie J. Andrews and Mary M. Talbot, eds., *All the World and Her Husband: Women in Twentieth Century Consumer Culture* (London: Cassell, 2000); Arjun Appadurai, "Introduction: Commodities and the Politics of Value," in *The Social Life of Things: Commodities in Cultural Perspective* (Cambridge: Cambridge University Press, 1986), 3–63; Pierre Bourdieu, *Distinction: A Social Critique of the Judgement of Taste*, trans. Richard Nice (London: Routledge and Kegan Paul, 1984); James G. Carrier, *Gifts and Commodities: Exchange and Western Capitalism since 1700* (London: Routledge, 1995); Mary Douglas and Baron Isherwood, *The World of Goods: Towards an Anthropology of Consumption* (New York: Norton, 1978); Paul Du Gay and Michael Pryke, *Cultural Economy: Cultural Analysis and Economic Life* (London: Sage, 2002); Ben Fine and Ellen Leopold, *The World of Consumption* (London: Routledge, 1993); Peter Jackson et al., *Commercial Cultures: Economies, Practices, Spaces* (London: Berg, 2000); Igor Kopytoff, "The Cultural Biography of Things: Commoditization as Process," in Appadurai, *The Social Life of Things*, 64–91; Celia Lury, *Consumer Culture* (New Brunswick: Rutgers University Press, 1996); Grant D. McCracken, *Culture and Consumption: New Approaches to the Symbolic Character of Consumer Goods and Activities* (Bloomington: Indiana University Press, 1986); Daniel Miller, *Material Culture and Mass Consumption* (Oxford: Blackwell, 1987); idem, ed., *Acknowledging Consumption: A Review of New Studies* (London: Routledge, 1995); Marshall Sahlins, *Culture and Practical Reason* (Chicago: University of Chicago Press, 1976); Don Slater, *Consumer Culture and Modernity* (Cambridge: Polity Press, 1997); Lynn Spigel, *Make Room for TV: Television and the Family Ideal in Postwar America* (Chicago: Chicago University Press, 1992); idem, *Welcome to the Dreamhouse: Popular Media and the Postwar Suburbs* (Durham: Duke University

Press, 2001); Rosalind Williams, *Dream Worlds: Mass Consumption in Late-nineteenth-Century France* (Berkeley: University of California Press, 1983).

3. For the notion of repertoire see Itamar Even-Zohar, "The Making of Culture Repertoire and The Role of Transfer," *Target* 9, 2 (1997): 373–381, and "Factors and Dependencies in Culture: A Revised Draft for Polysystem Culture Research," *Canadian Review of Comparative Literature* 24, 1 (1997): 15–34; Rakefet Sela-Sheffy, "Models and Habituses: Problems in the Idea of Cultural Repertoires," *Canadian Review of Comparative Literature* 24, 1 (1997): 35–47.

4. For this approach in magazine research see Marjorie Ferguson, *Forever Feminine: Women's Magazines and the Cult of Femininity* (Aldersho, UK: Gower, 1983); Janice Winship, *Inside Women's Magazines* (London: Pandora Press, 1987); Ros Ballaster et al., *Women's Worlds: Ideology, Femininity and the Woman's Magazine* (London: Macmillan, 1991); Hilary Radner, *Shopping Around: Feminine Culture and the Pursuit of Pleasure* (London: Routledge, 1995).

5. See Tamar Liebes and Elihu Katz, *The Export of Meaning: Cross-Cultural Readings of Dallas* (New York: Oxford University Press, 1990); Jostein Gripsrud, ed., *Television and Common Knowledge* (London: Routledge, 1999); Sonia Livingstone and Peter Lunt, *Talk on Television: Audience Participation and Public Debate* (London: Routledge, 1994); Horace Newcomb and Paul M. Hirsch, "Television as a 'Cultural Forum,'" in *Television: The Critical View*, Horace Newcomb, ed. (New York: Oxford University Press, 1994); 503–515, Roger Silverstone, ed., *Visions of Suburbia* (London: Routledge, 1997); Alison J. Clarke, *Tupperware: The Promise of Plastic in 1950s America* (Washington: Smithsonian Institution Press, 1999).

6. Janet Carsten and Stephen Hugh-Jones, *About the House: Levi-Strauss and Beyond* (Cambridge: Cambridge University Press, 1995); Donna Birdwell-Pheasant and Denise Lawrence-Zuniga, *House Life: Space, Place and Family in Europe* (Oxford: Berg, 1999); Irene Cieraad, *At Home: An Anthropology of Domestic Space* (Syracuse: Syracuse University Press, 1999); Tony Chapman and Jenny Hockey, *Ideal Homes? Social Change and Domestic Life* (London: Routledge, 1999), and Daniel Miller, ed., *Home Possessions: Material Culture Behind Closed Doors* (Oxford: Berg, 2001).

7. Ze'ev Posner, "The Flat and the Street in Israel as a Component of World Modelling" (M.A. Thesis, Tel Aviv University, 1998) (Hebrew); Rivkah Bar-Yosef, "Household Management in Two Types of Families in Israel: Applying an Organizational Model to Comparative Analysis," in *Families in Israel*, edited by Leah Shamgar-Handelman and Rivkah Bar-Yosef (Jerusalem: Academon, 1991), 169–196 (Hebrew); Daphna Birenbaum-Carmeli, *Tel Aviv North: The Making of a New Israeli Middle-Class* (Jerusalem: Magnes Press, 2000) (Hebrew).

8. Itamar Even-Zohar, "The Making of Culture Repertoire"; idem, "Factors and Dependencies in Culture"; Sonja Laden, "Middle-Class Matters, or, How to Keep Whites Whiter, Colors Brighter, and Blacks Beautiful," in *Critical Arts* 11: 1, 2 (1997): 120–141; idem, "Domesticity and Representations of Self and Reality in

la-'Isha, an Israeli Women's Weekly," in *Kesher* 28 (2000): 36–42 (Hebrew); idem " 'Making the Paper Speak Well,' or, the Pace of Change in Consumer Magazines for Black South Africans," *Poetics Today* 22, 2 (2001): 515–548; idem, "Magazine Matters: Toward a Cultural Economy of the South African (Print) Media," in *Media, Democracy and Renewal in Southern Africa*, Keyan Tomaselli and Hopeton Dunn, eds. (International Academic Publishers: Denver, 2000), 181–208; Gideon Toury, "Culture Planning and Translation," forthcoming in *Proceedings of the Vigo Conference "anovadores de nós-anosadores de vós,"* A. Alberto Alvarez et al., eds.

9. See Matthew Schneirov, *The Dream of a New Social Order: Popular Magazines in America, 1893–1914* (New York: Columbia University Press, 1994). See also, Richard Ohmann, *Selling Culture: Magazines, Markets, and Class at the Turn of the Century* (London: Verso, 1996); Michael Schudson, *Advertising, the Uneasy Persuasion: Its Dubious Impact on American Society* (USA: Basic Books, 1984); Williams, *Dream Worlds*.

10. See, e.g., Margaret Beetham, *A Magazine of Her Own? Domesticity and Desire in the Woman's Magazine, 1800-1914* (London: Routledge, 1996); Ferguson, *Forever Feminine*; Lori Anne Loeb, *Consuming Angels: Advertising and Victorian Women* (New York: Oxford University Press, 1994); Katherine Shevelow, *Women and Print Culture: The Construction of Femininity in the Early Periodical* (London: Routledge, 1989).

11. Michael Keren, "The Woman and Civil Society in Eretz Israel during the 1920s," *Kesher* 28 (2000): 36–42 (Hebrew); Hanna Herzog, "The Women's Press in Israel: An Arena for Reproduction or Challenge?" *Kesher* 28 (2000): 36–42 (Hebrew); Sonja Laden, "Domesticity and Representations of Self and Reality in *la-'Isha* . . . ."; Olga Cohen, "Models of Womanhood as Extracted from Hebrew Popular Magazines of the 1930s and 1940s in the Yishuv Culture in Palestine" (M.A. Thesis, Tel Aviv University. In preparation) (Hebrew).

12. See Schneirov, *The Dream of a New Social Order*; Scanlon, *Inarticulate Longings*; Beetham, *A Magazine of Her Own*; Ellen Gruber Garvey, *The Adman in the Parlor: Magazines and the Gendering of Consumer Culture, 1880s to 1910s* (Oxford: Oxford University Press, 1996) and Ohmann, *Selling Culture*.

13. Cohen, "Models of Womanhood."

14. See Keren, "The Woman and Civil Society in Eretz Israel," 28–35; Nava Cohen-Avigdor, "Female Politicians (vis-à-vis Male Politicians) in the Israeli Women's Press: Representations During Election Years 1959, 1977, 1996" (M.A. Thesis, Bar Ilan University, 1998), 360 (Hebrew).

15. In the late 1980s *la-'Isha*'s circulation figures stood at 89, 231. Dan Caspi and Yehiel Limor, *The Mediators: The Mass Media in Israel, 1949–1990* (Tel Aviv: Am Oved, 1992), 80–81 (Hebrew). Today's readership is estimated at approximately 600,000 (personal communication with current editor Orna Naner, July 2002).

16. Indeed, contrary to the accepted view of women's magazines as providing a typically intimate, personalized experience, located exclusively at home,

contemporary informal declarations by about forty Israeli readers of *la-'Isha* locate the reading of women's magazine in Israel in overtly public and semi-public spaces, such as, notably, doctors' and dentists' waiting rooms and hairdressing salons. Although these declarations should not be taken at face value, for they are clearly intended as demeaning, the fact remains that *la-'Isha* was (and still is) accessible, and frequently consumed by people specifically in public, rather than private, spaces.

17. See Beetham, *A Magazine of Her Own?*; Ferguson, *Forever Feminine*; Ohmann, *Selling Culture*; Nancy Walker, *Shaping our Mother's World: American Women's Magazines* (Mississippi: University of Mississippi Press, 2000).

18. *Ha-`Olam ha-ze* was initially titled *Tesh`a ba-èrev* (Nine in the Evening). It was renamed *ha-`Olam ha-ze* in 1946. Caspi and Limor, *The Mediators*, 78 (Hebrew).

19. Cohen, "Models of Womanhood."

20. Caspi and Limor, *The Mediators*, 78.

21. Zvi Elgat and David Paz, '*Isha (Woman) 2000: The Stories, the Dramas, the Style and the Beauty in Israel, 1948–2000* (Tel Aviv: Yedi`ot Akhronot, 1999), 15.

22. Elgat and Paz '*Isha (Woman)*, 15.

23. Ibid., 15.

24. Ibid., 15–16.

25. See Gershon Shafir and Yoav Peled, *Being Israeli: The Dynamics of Multiple Citizenship* (Cambridge: Cambridge University Press, 2002); and Assaf Razin and Efraim Sadka, *The Economy of Modern Israel: Malaise and Promise* (Chicago: Chicago University Press, 1993).

26. Don Slater and Fran Tonkiss, *Market Society: Markets and Modern Social Theory* (Cambridge: Polity, 2001).

27. Anat Helman, "The Development of Civil Society and Urban Culture in Tel Aviv During the 1920s and 1930s" (Ph.D. dissertation, Hebrew University, Jerusalem, 2000) (Hebrew); Oz Almog, *The Sabra: A Profile* (Tel Aviv: Am Oved, 2001) (Hebrew).

28. See, e.g., an item titled "On Improving *totzeret ha-'aretz*" (April 15, 1937) (Hebrew), penned by a women named Frieda, in which she complains (a) about the poor quality of woolen socks manufactured by the Israeli-owned plant "Lodzia," and (b) about the shopkeeper's response to her grievance whereby he suggested she buy imported socks in the future.

29. For other accounts of similar processes in England and the United States, some at different points in time, see Amy Beth Aronson, *Taking Liberties: Early American Women's Magazines and Readers* (New York: Praeger, 2002); Ferguson, *Forever Feminine*; David Paul Nord, "A Republican Literature: Magazine Reading and Readers in Late-Eighteenth-Century New York," in Cathy N. Davidson, ed., *Reading in America* (Baltimore: The Johns Hopkins University Press, 1989), 114–139; Scanlon, *Inarticulate Longings*; Walker, *Shaping our Mother's World*; Williams, *Dream Worlds*.

30. Ferguson, *Forever Feminine*, 18–22; Karal Ann Marling, *As Seen on TV: Visual Culture of Everyday Life in the 1950s* (Cambridge, MA: Harvard University Press, 1994), 202–240; Walker, *Shaping our Mother's World* 66–100.

31. Ferguson, *Forever Feminine*.

32. Ibid., Walker, *Shaping Our Mother's World*.

33. Bat-Oren, "Kedai la-da`at," inside page.

34. For a more detailed account of the methodology underlying magazine production see Laden, "Middle-Class Matters," 125.

35. Bat-Oren, "Kedai la-da`at," 66–67.

36. Ibid., 66–67.

37. Ibid., 65.

38. Ibid., 68–69.

39. Ibid., 70.

40. Shafir and Peled, *Being Israeli*; Razin and Sadka, *The Economy of Modern Israel*.

PART III

Markets, Consumption, and Inequality

CHAPTER 5

Consumption and the Place of the Economy in Society: "Reciprocity" and "Redistribution" in Markets for Houses and Household Durables in Republican Turkey

Ayşe Buğra

During the last two decades, many researchers have approached social change in Western developed countries by concentrating on processes of consumption. This observation pertains to different currents of analysis such as the historical studies of the advent of the consumer society in the eighteenth century,[1] Regulation School type of analyses investigating the transition from a Fordist to a post-Fordist accumulation regime,[2] and consumption-related research within cultural studies.[3] It does not seem possible, however, to make a parallel observation as far as the analyses of late industrialization are concerned. Notwithstanding certain important contributions that have mainly addressed problems such as the divergence of consumption activity from "genuine" need satisfaction or policy failures leading to parallel failures in the satisfaction of "basic needs,"[4] historical analysis of the relative significance of the roles played by states, markets and households in determining consumption patterns does not form an important part of development literature.[5]

Studies on Turkish modernization and development constitute a typical example of this generalized neglect of consumption as a significant aspect of economic development and social change. This constitutes a rather strange phenomenon since the transformation of household consumption patterns was in many ways central to the modernization project pursued by the Republican elite. Neglecting consumption has left, therefore, many interesting questions concerning the mechanisms that articulated macro- and micro-level historical change. This chapter attempts to explore one set of such questions, those that pertain to the sociopolitical mechanisms through which the individual is integrated in a society in transformation.

My concern lies, therefore, less in social history per se than in the political economy of consumption. The investigation of political and social determinants of individual consumption presented here is carried out with the idea that such an investigation could make an important contribution to the analysis of the ways in which "the economy is instituted" in modern Turkey. This formulation of the objective of the study reflects a Polanyite perspective whereby the economy is not reduced to a series of self-regulating markets, but viewed as a "process" instituted through the society-specific interaction among the principles of exchange, redistribution and reciprocity, with their accompanying institutional patterns. Polanyi investigates the "place of the economy" in traditional as well as modern societies by arguing that along with market exchange, principles of redistribution, made effective by the central state, and reciprocity, relying on kinship or community ties, also appear in different social arrangements designed both to prepare the institutional basis of the market and to complement it in the satisfaction of human needs.[6] Of these three principles, exchange and redistribution are formal in character while reciprocity is personal and informal. Relations of solidarity, trust and mutual responsibility that reflect the social positions of people involved characterize reciprocity in its difference from the impersonality and anonymity of market relations or state redistributive processes. In general, relations of reciprocity follow the family metaphor in their different manifestations among fellow-townsmen, neighbors, religious or ethnic community members, or within mafia-like organizations. The social context of reciprocity relations is not, however, limited to these examples. They could also permeate the realm of the market or the state, transforming the formal, rule-bound character of exchange or redistribution by the informality and personality of family-like relations. The principle of reciprocity could thus be used with reference to divers phenomena such as clientelism, cronyism, and corruption that might or might not involve organized crime networks, which political economists and social scientists have recently found themselves in a position to address in the study of contemporary societies. It could of course be suggested that such an approach turns reciprocity into a "residual" category under which all sorts of deviant behavior could be grouped. Still, the three-fold classification of Polanyi keeps its superiority over currently very influential economistic approaches where the principle of self-interest, the motive force behind exchange relations, is evoked to analyze practically all types of social relations. Moreover, as long as one remains sensitive to many divers forms that they can take, the emphasis of reciprocity relations would enable the researcher to go beyond the simplistic dichotomy between the state and the market in a more productively complex analysis of the way in which the economy is instituted. It is with these methodological concerns in mind that the following discussion of political economy of consumption in Turkey highlights the centrality of reciprocity in the place of the economy in Turkish society.

Demonstrating how the functioning of the market and the role the state plays in the economy could come to reflect the logic of reciprocity calls for a comparative approach. The discussion presented in this chapter thus draws attention to those factors that distinguish the social context of consumption activity in Western developed countries from the ways in which this context is shaped in Turkey. Insights drawn on the Regulation School of Analysis appears to be particularly useful for the purposes of such a comparative evaluation. This school emphasizes the specific articulation of the dynamics of production and consumption as the defining feature of a given "regime of accumulation." It highlights, consequently, that a regime of accumulation cannot be reduced to labor process, but is also characterized by the particular structure of effective demand, that validates the trends in productive process, and by the nature of the relationship between market and non-market forms of economic activity in society. Starting with this point, several writers have shown that one of the most significant aspects of the post–Second World War development in Western developed countries have been the management of household consumption in a way to assure the participation of the individual in social life.[7]

Here consumption refers to all those activities, situated outside the realm of production, through which the individual seeks to participate in and to be a part of the society in which (s)he is situated. The boundaries of consumption activity in this sense are drawn by socially determined needs and forms of need satisfaction, and the realm of consumption is not limited to the realm of the market. Social rights that reflect the nature of state redistributive processes as well as networks of personal relations, too, are often effective in determining the coordinates of the livelihood of the individual and giving content to the meaning of citizenship in a particular society. The Regulation School of Analysis highlights that in all Western economies, it was the crucial role played by the state that has assured socioeconomic integration. The integration of people in society through consumption activity has, in other words, been an important aspect of social policy especially visible in the sectors of housing and household durables.

In republican Turkey, too, debates around the nature of "modern Turkish house" have been a significant aspect of the ideological atmosphere of early modernization attempts. These debates pertaining to houses and household appliances were all the more significant because they were directly related to the changing place of women in society, where the historical dynamics of modernization was thought to manifest itself more clearly than anywhere else. In the Turkish context, however, personal relations of trust, solidarity, and mutual responsibility in their informal character seem to have assumed an important role as mechanisms through which consumer needs are determined and satisfied, and the individual is integrated in society.

This contrast is especially clear in the realm of housing. In all Western developed economies, including the most typically liberal ones such as the United States, the need for shelter has been regarded as a basic one whose satisfaction could not be left entirely to the market. The state has appeared, therefore, as a crucial actor in shaping consumption in this area through redistributive policies of divers types. These policies range from the mortgage system, which is central to the functioning of the American financial sector, to the provision of low-cost social housing to the working population in many European countries. While the state presence in Turkish economy has in general been much more significant than in these Western economies, we don't see the redistributive role of the state as a crucial factor shaping the housing sector in Turkey. Although the transformation of residential architecture has significantly figured in the overall social modernization project of the Republican State, the latter has largely abstained from the management of consumption activity in this area. Through massive waves of rural–urban migration that mark the socioeconomic developments of especially the post–Second World War era, housing sector has been largely organized on the basis of reciprocity relations involving divers segments of the population. Typical of the irregular housing market in Turkey as elsewhere, reciprocity networks have in time permeated both the relations of homeowners from all classes and real estate developers with the state authority and the urban housing market in its totality. Housing sector as a whole has consequently come to be typified by *gecekondu*, Turkish term for irregular settlements.

As to the consumer durables sector, both in Western developed countries and in late industrializing countries such as Turkey, they constitute the most important category of mass consumer products in the twentieth century.

Historical studies suggest that in the development of this sector in most Western countries, the state has played a crucial role in two different ways: first, by contributing to the stability of employment and income through minimum wage laws and different social security provisions and, second, by setting the legal basis of consumer credit practices. In Turkey, the sector has benefited from protectionist trade measures in the form of high tariff and nontariff barriers. There were other types of state-provided support to the producers, such as subsidized credit or low-cost inputs from State Economic Enterprises. However, consumption management measures have been largely absent and networks of sales agents, formed by the producers, compensated their absence. Within these networks, relations among producers, sales agents and customers have often assumed a personal and informal character deviating, thus, from the principle of exchange and the market pattern. Largely functioning on the basis of the principle of reciprocity, these networks have constituted a deliberate response to the challenge of market formation by substituting for the role of redistributive policies that had, in the United States and in Europe, contributed to the expansion of mass consumer markets and to a certain standardization of household consumption patterns.

The following comparative evaluation of the post–Second World War developments in consumption patterns in housing and household durables sectors attempts to highlight, therefore, a certain asymmetry in the relative significance of the principles of reciprocity and redistribution in defining the nature of consumption activity in Turkey and in Western developed countries. I believe that this asymmetry is indicative of rather interesting societal differences that might extend beyond the realm of household consumption in particular to the place of the economy in society in general.

Reciprocity as the Organizing Principle of the Economy of Housing in Turkey

I have defined consumption as an activity whereby the individual attempts to participate in social life. It is an activity, therefore, which reflects both the nature of the society and the place of the individual in it. In Turkey, as in many other countries, this is nowhere as clear as it is in the housing sector. In Mona Russell's contribution to this volume, we find an excellent illustration of how "the ideal vision of the home" in the nineteenth- and early twentieth-century Egypt provides a lens for studying the social transformations taking place in Egyptian society in that period. A very close parallel is found in the centrality of debates around the nature of "the modern Turkish house" to the early Republican attempts at social modernization. In these debates, considerable significance was attached to the role of the architect in the realization of the modernization project. As a contributor to one of the first professional journals of architecture published in the country put it, the architect should not be considered as

> Someone who builds us a shelter against rain or sun, but as a thinker who gives us intellectual guidance in our domestic life. As he takes care of the outside appearance of our house, he also takes care of the inside, perhaps in a more important manner. We can no longer fill our house with ready-made furniture haphazardly purchased from a store. It is no longer possible for us to imitate the Europeans. This is why I am trying to describe to you, not the house of a German or French family, but our very own house.[8]

While such attempts to describe "our very own house" were going on, it was also recognized that planned intervention of the state would be required to organize the sector along socially desirable lines. The need for such intervention was especially strong since the state employees, who had settled in the new republican capital Ankara, then a small and backward central Anatolian town, had considerable difficulties to find adequate housing for their families. Hence, the state had to take action to satisfy the need for residential housing in Ankara and a number of cooperatives engaged in building residences for state employees were organized and regulated with government initiative. However, government policy in this area seems to have been designed to serve a wider objective on the basis of the assumption that the new residential patterns adopted by bureaucrats and civil servants would eventually filter through the rest of the population across different social groups and geographic regions. It was expected, in other words, that the houses built for the new Republican elite would set the pattern for the country as a whole.

It could be seen that the initiatives taken in this direction have sometimes extended beyond Ankara, to different Anatolian provinces, to the Eastern ones in particular. For example, in an official report on this subject, we find the following passage:

> According to the directives of our national leader,[9] the government is planning to undertake the construction of 5,000 residences for its employees in the eastern provinces. Fifty-five of these residences would be for the first class bureaucratic cadres and have four rooms, 1,000 of them would be for the second class cadres and have three rooms, and there will

be 3,500 residences with two rooms for the third class. This constitutes a major undertaking which is not only aimed at contributing to the supply of appropriate housing for civil servants, but is also designed to present an example to be emulated by the private residential construction activities in the future.[10]

In the 1940s, when this report was written, the government in power had, for the first time, included the housing problem in its program among other social policy matters. As it was indicated in the program, the government intended to continue "residential construction in Eastern provinces, both to satisfy the housing needs of civil servants and to improve the urban architecture in the region by presenting the local people with good examples of modern residential buildings."[11] The architects and city planners of the era, that were critical of this particular policy orientation, which consisted mainly in the construction of residences for civil servants, have repeatedly warned political authorities that the popular housing problem would reach formidable dimensions in the future if proper precaution was not taken.[12] It could not be said that republican governments in power have heeded these warnings. Not only there was no public funding allocated to social housing projects but attempts to regulate urban construction activity, too, remained very limited. In fact, bureaucrats, that were supposed to regulate the housing sector in conformity with the norms of modern urban planning practices, were the first ones to trespass rules for speculative purposes. These bureaucrats have extensively mobilized their political relations to change or circumvent policy decisions to enhance the value of their newly acquired private plots of land and, later, to bypass building regulations to increase the allowed area of construction on these plots. The initial concessions made in these areas have led the way for explicit government decrees that allowed, first, the construction of commercial buildings in residential areas and, subsequently, the vertical growth of buildings through the addition of new floors. Those who were in a position to introduce and implement social housing projects were, in other words, too busy accumulating speculative fortunes to think about matters relating to popular housing.[13] The state's Real Estate Credit Bank founded to provide subsidized credit for home ownership, too, has ended up subsidizing middle-class residential projects in Ankara and Istanbul. Through most of Republican era, the role of the state in the area of low-income residential sector has been confined to the implementation of rent controls which, to the extent that they were not circumvented by landlords easily exploiting the prevailing sellers' market, have also contributed to the shortage of housing.[14]

The fact that the Turkish state does not appear as an important actor in regulating consumption activity in the housing sector has not, however, led to a situation where the satisfaction of this vital social need has been left to the market. Instead, the role that the welfare state has played in developed Western countries was largely assumed by irregular settlements. The development of irregular settlements has taken place outside the formal, legally bound channels of exchange and redistribution. In ways that are highly reminiscent of Palestinian house-building strategies that Tania Forte (chapter 6) discusses in her contribution to the present volume, kinship ties, as well as the relations of fellow-townsmenship, have been mobilized within reciprocity networks and have led to the emergence and increasing importance of a particular type of self-help housing, *gecekondu*.

While such irregular settlements can be found in many late industrializing countries of Asia, Africa, and Latin America, in Turkey, as most probably elsewhere, they also have certain characteristics that reflect society-specific forms of interaction among the principles of exchange, redistribution and reciprocity. In this regard, it is important to note that in Turkey, the *gecekondu* pattern has unfolded against the background set by the nature of the role of the state in Turkish society. In Turkey, the state has always been present in a significant way, although its role in economy and society could hardly be defined with reference to the logic of redistribution. This is manifested, for example, in the reciprocity networks mobilized by the early Republican elite in their attempts to realize speculative gain from real estate ownership in Ankara, which have set the pattern for the subsequent developments in the housing sector. Through these developments, "access through personal relations," rather than market emanated purchasing power, has determined the possibility of home ownership.

The nature of the role Turkish state plays in economy and society is also reflected in the significance of urban land tenure patterns in giving shape to the *gecekondu* pattern. Since the role of formal redistributive measures have been very limited in meeting urban residential requirements, the urban poor could only aspire to the ownership of a *gecekondu* with the help of their families and friends, who would provide the necessary resources including information, finance, and manpower. However, unless the conditions of access to land were not hopelessly difficult, these resources would be of little use. In Turkey, they could be mobilized effectively because the state has exercised an important influence in the housing sector through the large supplies of public land in and around the cities. Neither privatized nor used for social housing projects, these vacant public plots have begun to be treated as "commons" and appropriated by successive waves of immigrants from rural areas.

In fact, it would not be unfounded to suggest that if the urban land tenure patterns had been different and private property in urban land had been more significant than it was, the development of irregular settlements could have hardly reached its current dimensions.[15]

Another characteristic of the urban land tenure patterns in Turkey has been the ambiguities surrounding the entitlements to the peripheral agricultural land in the vicinity of big cities. The unpredicted expansion of big urban centers toward the periphery has eventually made these plots an integral part of the urban geography without changing their legal status. They have remained under the jurisdiction of village authorities and different rules applied with regards to their ownership and development. Residential buildings on unauthorized subdivisions of such rural land that have appeared without regular titles and construction permits today constitute the most important type of irregular housing in Turkey.[16]

The importance of public property in land has given Turkish governments ample opportunity to opt for clientelistic politics, by exchanging entitlements to invaded public property for votes. This reciprocal exchange of votes for title deeds has become all the more important given the confusion concerning the ownership and use of peripheral agricultural land. Through the clientelistic politics, which politicians preferred to the less rewarding pursuit of a systematic social housing policy, most squatters have acquired property titles to their houses through successive amnesty laws. In this process, irregular housing has ceased to be the site of the precarious existence of marginal segments of the population and has become an undeniable aspect of the residential sector with an important commercial potential that could be entrepreneurially exploited.

The commercialization of *gecekondu* has taken place along with a parallel process whereby middle-class housing market has begun to lose its formal, rule-bound character. Irregular practices involving the violation of building regulations or dubious methods of acquiring of building permits have become a permanent features of the way housing market, in its totality, functions. Housing sector has been gradually permeated by networks of reciprocity that involve exchange of economic or political favors among landowners, developers and politicians at all levels. In other words, the *gecekondu* pattern has become central to the urban land market in general. This is the ironical situation reached through a series of historical developments that had begun with deliberate attempts at the modernization of residential housing patterns. While the early Republican elite hoped to see the modern houses they built for civil servants being emulated by the public at large, what was emulated was the way state employees have mobilized personal relations to change or to circumvent rules and regulations for speculative purposes. In the absence of formal redistributive policy measures regulating consumption activity in the housing sector, relations of reciprocity have marked the sector as a whole.

Non-Market Determinants of the Market for Consumer Durables

Even in those years when they were not even remotely accessible to the overwhelming majority of Turkish consumers, household appliances had a significant place in the early Republican imagination of social modernization. It is possible to find, in women's magazines of the 1930s and 1940s, many debates around consumer durables as basic need items. As important bearers of a modernizing mission, these magazines have insistently presented these items to their female readers in terms such as the following:

> Today, a decent set of bedroom furniture costs at least one thousand liras. Even those who could buy such a set by paying cash would be much more comfortable and much happier if they decide to use that sum of money in an alternative way, to purchase either a sewing machine, a gas oven, a radio or even a refrigerator, any one of these items that are indispensable for a modern house.[17]

Different issues of this magazine and others have included, through the 1930s and the 1940s, information and advise of the same type, with accompanying pictures of Western—mainly American—households equipped with modern household appliances. The way these pictures were presented sometimes had the quality of science fiction since it incorporated an imagination that was much beyond not only the Turkish but also American realities. For example, a picture, obviously representing an industrial kitchen with a female worker feeding dirty dishes into a dishwasher-type machine, is presented as "the lady of the house washing dishes in a dishwasher" in a schoolbook on home economics printed in 1938. People, or rather women, were thus presented with modern fantasies or "fetishes feeding the contemporary imagination" as a historian of Turkish republican architecture puts it.[18] In fact, a few statistics could easily reveal the magnitude of the gap between fantasy and reality: In the early 1940s, in Turkey, only 123 of 461 urban district centers, twenty-one of 940 rural town centers, and eight of over 35,000 villages had electricity.[19]

A relative narrowing of the gap between fantasy and reality begins in the late 1950s with the establishment of a domestic enterprise producing washing machines and refrigerators. We see the women's magazines of the period reporting this important development, along with interviews with housewives expressing themselves in terms such as the following:

It is my greatest wish to have a washing machine and a refrigerator, which I regard as indispensable items for my house. I am very happy that these items are now being produced in our country and I have already begun saving to purchase them. I sincerely hope that they will be sold at prices accessible to all housewives.[20]

In those early years of local production of household durables, the number of housewives who could afford them seems to be extremely limited. In a country of approximately twenty-eight million people, less than 2,000 washing machines were sold in 1959 and the number of refrigerators sold in 1960 was less than 1,500.[21] Nevertheless, the growth in sales was very rapid. In less than three decades, mass consumption markets on a national scale could be established not only in washing machines and refrigerators, but also in other household durables such as vacuum cleaners and kitchen ovens and, later, in television and music sets, dishwashers, and smaller electrical appliances.

The conditions under which this market expansion has taken place were considerably different from those characterizing the development of the household durables sector in Western industrialized countries in the post–Second World War period. On the one hand, the import substituting industrialization policy in implementation in Turkey provided a very favorable environment for producers who did not have to compete with foreign imports. Domestic competition, too, was very limited. Although the first Turkish enterprise that entered the field, Arçelik, was soon followed by a second producer, until the late 1980s the market for most consumer durables exhibited the characteristics of a typical duopoly. Producers also benefited from cheap inputs supplied by State Economic Enterprises and low credit costs that resulted from interest rate controls.

On the other hand, there was no state support provided to the consumer. While the retail prices of domestically produced consumer durables were considerably lower than those of imported ones, which were simply inaccessible to the majority, they were still very high in comparison to the incomes of wage and salary earners.[22] Moreover, the latter were a minority in the labor force, which mainly consisted in small producers, mostly situated in agriculture. Along with low and unstable incomes, basic infrastructure, too, was very inadequately developed. Electrification at the national level was far from being complete and percentage of households, which did not have city water, was considerable.[23] Under these circumstances, the increased sales of household durables required special mechanisms—both to create the need for the products and to facilitate payments for them. At the prevailing low levels of income, the only possible means of market expansion appeared to be sales by installments, which were extremely difficult to organize in the absence of appropriate legislation and institutions.

In fact, perhaps a more important marketing difficulty facing the producers was the institutional vacuum in the realm of consumer credit. In most developed Western countries, the legislation regulating consumer credit came earlier in history than standard welfare state provisions. Hire-purchase or *vente-a-tempérament* type arrangements were found in different countries in the nineteenth-century, as in the case of furniture and household equipment sales through installments. In general, there seems to be a very clear relationship between generalized ownership of household durables, increase in installment credit, and state legislation regularizing the legal framework of consumer credit arrangements. It is not surprising, therefore, that the earliest and most elaborate legislation in this area was introduced in the United States to spread, eventually, to different European countries, significantly contributing to the expansion of the markets for consumer durables.[24]

In Turkey, consumer credit arrangements have been regularized by two clauses of the Loan Act, which was enacted in 1926 on the basis of an early twentieth-century amendment (1911) of a nineteenth-century Swiss law (1880). After 1911, the Swiss law was amended six times while the Turkish Act has remained the same. In sales by installments, the protection that the clauses in question provided to the seller was limited to sales contracts signed in the presence of a notary general in the residence of the purchaser. While the necessary bureaucratic formalities introduced enormous transaction costs in the exchange process, the chances of the seller getting a product back from a customer who did not honor the terms of the contract would be very slim—possible only if the product has not, in the meantime, been sold to an innocent third party. With these limitations, the Loan Act alone could never constitute a solid institutional basis for the development of the mass consumer market for household durables. Moreover, Turkish legislation prevented the formation of credit agencies specializing in the provision of consumer credit. While there were several attempts made to allow first the Chambers then the Central Bank to form and manage credit agencies specialized in consumer credit, no proper revision of the Bank Act was undertaken in this direction.

This institutional vacuum has been a constant source of criticism voiced by the managers of the Arçelik company who systematically pressured the government to revise the Loan Act to provide more protection to the seller and to modify the Bank Act in a way to allow for the establishment of financial intermediaries specialized in consumer credit. The producers were in fact demanding that the state play a more active role in market formation. Since these demands were not satisfied, they had to devise an alternative institutional mechanism to substitute for the role that the state had assumed in the development of mass consumption markets in Western countries.

The institutional mechanism in question took the form of a nationwide network of dealers operating on the basis of trust and loyalty rather than in conformity with the impersonal, formal character of market exchange between anonymous individuals.[25] The number of dealers in the Arçelik network reached 2,400 in the first couple of years after the establishment of the company. However, most of them were soon eliminated with 1,000 of them remaining as exclusive agents of the manufacturer. These exclusive agents, whose number has reached over 3,000 in the 1990s, were chosen among those people, often small merchants, respected and trusted in their locality. They have remained in the network to the extent that their identification with and sense of belonging to the company was strong enough. The ties of the retailers with the company, in their turn, were systematically reinforced by deliberate efforts undertaken by the company management. The idea that an Arçelik dealership is not a simple business venture but a long-term commitment to a respectable family was in many ways central to the efforts in question. The family metaphor was, in other words, the central component of the organizing rhetoric of the Arçelik marketing network. To maintain the respectability of the "family," sales agents were not only chosen carefully, but were also supervised closely. In this regard, instability of family life or the emergence of a drinking problem could constitute sufficient reasons for the termination of a dealer's contract. Less than perfect loyalty to the producer, being suspected of negotiating with the rival company for a higher sales commission for example, would also lead to a loss of dealership. In the meantime, several measures were taken to strengthen family ties. Arçelik was only one of the many ventures of the Koç Group of Companies. However, Vehbi Koç, the founding president of the Group, by far the richest and the most famous of Turkish businessmen of the republican era, was personally involved in the relations of Arçelik with its dealers. Annual meetings, where hundreds of dealers, and their families, came together in luxurious hotels located in different parts of the country, constitute an important aspect of these relations. Until his death a few years ago, Vehbi Koç himself hosted these meetings, which were reported in the Arçelik monthly bulletin with the aid of recurrent formulas such as "The largest and the most distinguished family of Turkey came together last month" or "The head of the family, Vehbi Koç, was once more among the old dealers."[26]

This emphasis placed on trust and loyalty among "Arçelik family" members could be understood with reference to the role the network of dealers has assumed in the household durables market. In the latter, both marketing and finance have exclusively relied on the dealers who were in a position to bring together the producer and the consumer in an entirely trust-based system. In an environment where consumer durables were entirely new consumer goods, it was the dealer, as a respected and trusted local figure, who convinced the customer that they were indispensable and affordable items. In a neighborly way, the dealer presented the potential purchasers with user instructions and payment plans that would not upset the family budget. In the institutional vacuum in which sales by installments had to take place, trust between the retailer and the customer had a vital role. Where the customer did not honor the sales contract, the retailer undertook the resulting risk, in its totality. The latter, often a small merchant without significant capital resources, was in a position to calculate and avoid risks very carefully for which he relied, almost exclusively, on the strength of his personal relations with the customer.

As to the relationship between the retailer and the producer, it was also a flexible one that did not always conform to the logic of market exchange. For example, in the second half of the 1970s, when Arçelik was faced with serious financial difficulties because of the nationwide strikes that paralyzed the manufacturing sector as a whole, it could remain afloat only thanks to the support of its retailers. These small merchants of very limited means have agreed to extend loans to the producer solely against the promise of deliveries of merchandise upon the termination of the strike activity. Given the extreme uncertainty prevailing in the sector and in the economy as a whole, the credit extended to Arçelik was an act of solidarity that could hardly be explained on the basis of sheer economic rationality characteristic of the impersonality and anonymity of the market relation. In a similar vein, during the severe economic crisis of 1994, Arçelik could only pay its dollar-denominated debts by borrowing from its dealers. This generosity was later reciprocated by the manufacturing company by keeping the wholesale prices constant although they were eroded in dollar terms as a result of the loss of value of Turkish currency.

These examples do not, of course, show that economic interests have not significantly figured in the functioning of the Arçelik sales network. They point, rather, at recognition of mutuality of interest, which has led different members of the network to support each other to assure the long-term survival of the system. Given the limitations of both purchasing power and infrastructure and, especially, the fragility of the formal legislative framework in which the market has operated, it would indeed have been impossible for the consumer durables sector to survive had short-term material interest maximization at each transaction been the behavioral norm. In Turkey, as elsewhere, market formation was neither spontaneous nor entirely based on the motive of short-term economic gain. However, the market was not formed and developed with the aid of state redistributive processes as in Western industrialized countries, but on the basis of personal and informal principle of reciprocity. This particular non-market mechanism of market formation was distinctly non-modern in the sense that relations of personal trust, typical of traditional societies, replaced the anonymity and impersonality of market exchange and state redistributive processes, which characterize modern market societies. The spatial organization of local dealers, too, had little to do with the images of modernity presented to the Turkish public within the ideological atmosphere of early republican modernization efforts. With their cluttered and somewhat messy appearance, these retail outlets reminded one more of the old-style dry goods stores characteristic of early American towns than of the pictures of modern kitchens printed in Turkish women's magazines or schoolbooks of home economics in the 1930s and 1940s.

In Lieu of Conclusion: Institutional Legacy in a Changing Environment

Transformation of private lifestyles is an important aspect of Turkish modernization project. Therefore, the changing character of houses and household furniture naturally appears as an integral component of this project undertaken by the republican state. Turkish state did not, however, play the deliberate market-forming role in these areas that were significantly shaped by the logic of redistribution through the Fordist industrial development of Western countries. Changing consumption patterns, which have accompanied Turkish modernization, were more attributable to the role of reciprocity networks in their society-specific forms of interaction with redistributive processes. This, I have argued, is an important historical factor, which defines the place of the economy in Turkish society in its difference from the way economy is instituted in North American and European societies. The difference in question, in its turn, is indicative of the interplay between the modern and the traditional in defining the twentieth-century setting of late industrialization in countries such as Turkey. It is possible to argue, therefore, that Turkish modernization has taken place in a way to challenge the neat distinction between the modern and the traditional that the founders of the Turkish republic had in mind.

Recently, the role of the state both in forming the market and in complementing it in the satisfaction of consumer needs has undergone certain changes with the societal transformations, often discussed with reference to the logic of post-Fordism. In this regard, declining importance of mass consumption markets for standardized products and the parallel retreat of the state in its consumption management role significantly figure in the recent historical dynamics of developed Western societies. Somewhat anachronistically, the Turkish state has begun to play a more significant role in this period. In the realm of housing, the Mass Housing and Investment Administration (MHIA) that was founded in 1984 soon became the largest housing finance agency in the country. In less than a decade, the total amount of residential construction units it financed have exceeded number of formal housing units in the post–Second World War era.[27] Not all of the residential units financed by the MHIA were designed to satisfy the needs of the urban poor. Yet, the government made serious attempts to channel funds toward popular housing projects, with special emphasis placed on the *gecekondu* problem.

This change in policy orientation was in large part related to two factors. First, environmental pressures of unplanned urban growth have reached unsustainable dimensions and it was no longer possible to overlook the deterioration of the natural and social fabric of the city. Second, with the increasing feeling of insecurity and discomfort associated with the deterioration of urban geography and the inadequacy of municipal services, middle-class residential patterns have begun to change. As the well-to-do families began to move from the center to the periphery of the big cities, peripheral urban land, hitherto available for the development of irregular settlements, has begun to be interesting for developers catering to middle- and upper-middle class demand for housing. Conflicts of interest around the contested plots in the vicinity of big cities have thus become instrumental in the government's attempts to assume a new role in the area of popular housing. Nevertheless, the legacy of long-lasting relations of reciprocity among politicians at all levels, squatters, and real estate developers is still strong enough to block the policy measures directed at the de-personalization and formalization of the role of the state in the housing sector. The fact that these measures have still not culminated in the emergence of redistribution and exchange as the dominant principles of behavior in the residential housing sector and the disappearance of *gecekondu* as the typical form of popular housing is indicative of the strong legacy of past historical trends.[28]

Albeit in another fashion, the legacy of the past is also visible in the consumer durables sector. As in the case of residential housing, in this sector, too, the post-1980 changes in the economic environment have called for certain renovations in organizational structures. Two factors appear to be particularly important in this regard: first, double digit inflation, established as a permanent character of the Turkish economy, has constituted an increasingly strong pressure on the mechanism of consumer credit administrated by dealers on a highly fragile basis, especially after the liberalization and the consequent rise of interest rates; second, the liberalization of the trade regime in the 1980s generated serious fears about the disruption of the duopoly situation characterizing the industry. In this new environment, some of the managers and consultants of Arçelik argued for the necessity of replacing the old sales network by a new, modern marketing system. The company also intensified the pressure for legislative changes of the Loan Act and the Bank Act. These pressures culminated in the revision of the Bank Act in 1994 and the foundation of Koç Finans, a specialized financial institution extending credit to customers of Arçelik without the intermediation of the dealer. Along with this modern credit institution, several modern outlets with standardized spatial arrangements for product display and demonstration have also made their appearance in the Arçelik sales network.

An outside observer of the contrast between the hygienic, laboratory-like atmosphere of these new outlets and modern Koç Finans bureaus on the one hand and the strictly "non-modern" appearance of old-style dealerships on the other could easily conclude that the latter form part of an archaic system bound to leave its place to a new system, which conforms to the logic of the exchange relationship as it prevails in all modern market societies. Behind the new facade, however, it was the old marketing network that has assured the smooth functioning of the retail system both by providing Koç Finans otherwise nonavailable information on consumer reliability and by acting as personal guarantors to their customers. Information provision and risk-sharing has remained, in other words, the responsibility of dealers, which they could undertake by virtue of their personal relations with their customers. In fact, the significance of this highly unequal manner of responsibility-sharing between the modern credit bureau affiliated with the manufacturer and the retailers was clearly revealed by the difficulties faced by a similar financial intermediary established by Arçelik's competitor, also in 1994. This competitor had previously formed a joint venture with two foreign companies, Siemens and Bosh, and proceeded to dismantle its own marketing network modeled after Arçelik's. Hence, its credit bureau had to operate in the absence of a reliable network of dealers, which has proven to be a serious impediment undermining the competitive position of the company against Arçelik. As the new institutional framework of consumer credit significantly relied on informal and personal relations characterizing the functioning of the old sales network, new outlets, too, were marked by similar relationships. The owners and/or managers of these outlets were often relatives of old dealers in the system and, just like the latter, they often seemed to internalize the family rhetoric systematically cultivated by the company.[29]

Housing and household durables sectors, discussed here as historical cases, have already begun to be altered by historical dynamics of transformation. Nevertheless, these cases in question have more than mere historical interest since they also show how socioeconomic change could be blocked or conditioned by past trends. Even if the relations of reciprocity that have long characterized popular housing and mass consumption of household durables are bound to give way to some combination of exchange and redistribution, the processes through which the new pattern would emerge are more than likely to be shaped by the legacy of the past.

I believe, however, that the significance of the principle of reciprocity lies beyond the cases discussed here or even beyond the realm of consumption as a whole. By highlighting a situation where the deeply felt economic presence of the state merely hides the centrality of reciprocity networks to the way the individual is integrated in society as a consumer, the cases in question provide an important insight into the analysis of the place of the economy in Turkish society. They suggest, in other words, that to understand the way in which economy is instituted in Turkey, we might have to look at neither the state nor the market but at "family-like" relations in their informal character, which tend to permeate the formal and legally bound contexts of these two institutions, reshaping them to their own image.

Notes

1. Neil McKendrick, John Brewer and J.H. Plumb, *The Birth of a Consumer Society: The Commercialization of Eighteenth-Century England* (Bloomington: Indiana University Press, 1982); Colin Campbell, *The Romantic Ethic and the Spirit of Modern Consumerism* (Oxford: Blackwell, 1987); John Brewer and Roy Porter, *Consumption and the World of Goods* (London: Routledge, 1993).
2. Michel Aglietta, *A Theory of Capitalist Regulation: The US Experience*, trans. David Fernbach (London: New Left Books, 1979); Robert Boyer, "The Eighties: The Search for Alternatives to Fordism" in Bob Jessop et al., eds., *The Politics of*

Flexibility: Restructuring State and Industry in Britain, Germany, and Scandinavia (Aldershot, UK: Edward Elgar, 1991), 106–132; Martyn Lee, *Consumer Society Reborn: The Cultural Politics of Consumption* (London: Routledge, 1993); Alain Lipietz, *Mirages and Miracles: The Crises of Global Fordism*, trans. David Macey (London: Verso, 1987).

3. David Harvey, *The Condition of Postmodernity: An Enquiry into the Origins of Cultural Change* (Oxford: Blackwell, 1989); Scott Lash and John Urry, *Economies of Signs and Space* (London: Sage, 1994).

4. Sam Cole and Ian Miles, *Worlds Apart: Technology and North–South Relations in the Global Economy* (Brighton: Wheatsheaf Books, 1984); David Felix, "Interrelations between Consumption, Economic Growth, and Income Distribution in Latin America since 1800" in Henri Baudet and Henk van de Meulen, eds., *Consumer Behavior and Growth in the Modern Economy* (London: Croom Helm, 1982), 133–177; Frances Stewart, *Planning to Meet Basic Needs* (London: MacMillan, 1985).

5. For a discussion of the analytical and political problems associated with the attempts to distinguish between "genuine" and "false" or "basic" and "nonbasic" needs, see Ayşe Buğra and G. Irzık, "Human Needs, Consumption and Social Policy," *Economics and Philosophy*, 15 (1999): 187–208.

6. Karl Polanyi, *The Great Transformation* (Boston: Beacon Press, 1944); Karl Polanyi, "Economy as Instituted Process" in Karl Polanyi, Conard M. Arensberg and Harry W. Pearson, eds., *Trade and Market in the Early Empires: Economies in History and Theory* (Chicago: Gateway, 1957), 243–270; Karl Polanyi, *The Livelihood of Man* (New York: Academic Press, 1977).

7. Aglietta, *A Theory of Capitalist* Regulation; Boyer, "The Eightees: The Search for Alternatives to Fordism."

8. Mimar Abdullah Ziya, "Binanın içinde Mimar," *Mimar*, 1, 1 (1931): 13–17.

9. The reference is to İsmet İnönü, who became the head of the state after Mustafa Kemal's death.

10. Anonymus, *Arkitekt*, 14, 11/12 (1944): 278–283.

11. Zeki Sayar, "Mesken Davası I & II," *Arkitekt*, 16, 5/6 (1946): 171–172 and 16, 7–8 (1946): 149–150.

12. Abidin Mortaş, "Ankara'da Mesken Meselesi," *Arkitekt*, 13, 11/12 (1943): 239–240. But more than anyone else, it was Zeki Sayar who has systematically voiced his opinion and offered policy advice on matters pertaining to urban planning through the three decades following the foundation of the Republic. See, e.g., "Mesken Davası I & II"; "İnşaat Kalfaları Problemi," *Arkitekt*, 17, 9/10 (1947): 199–200; "I. Türk Yapı Kongresinden Bekledikerimiz," *Arkitekt*, 18, 1/2 (1948): 19–34; "1952 Mesken Faaliyeti Nasıl Olacak?" *Arkitekt*, 21, 11/12 (1951): 205, 232; "Bizde Mesken Finansmanı," *Arkitekt*, 22, 9/10 (1952): 253–254.

13. Falih Rıfkı Atay, *Çankaya: Atatürk devri hatıraları* (Istanbul: Dünya Yayınları, 1958); Sezai Göksu, "Yenişehir'de Bir İmar Öyküsü," İlhan Tekeli, der., *Kent,*

Planlama, Politika ve Sanat: Tarık Okyay Anısına Yazılar (Ankara: ODTÜ Mimarlık Fakültesi, 1994), 257–276.

14. G. Pulat, *Dar Gelirlilerin Konut Sorunu ve Soruna Mekansal Çözüm Arayışları* (Ankara: Kent Koop, 1992) presents a full survey of the housing policy in the republican period and the relative effectiveness of the instruments used. See also ılhan Tekeli, "70 Yıl İçinde Türkiye'de Konut Sorununa Nasıl Çözüm Arandı?" *Konut Araştırmaları Sempozyumu*, Konut Araşırmaları Dizisi I (Ankara: Toplu Konut İdaresi Başkanlığı, 1995), 1–10; Ruşen Keleş, "Housing Policy in Turkey" in Gil Shidlo, ed., *Housing Policy in Developing Countries* (London: Routledge, 1990), 140–172.

15. Buğra, "The Immoral Economy of Housing in Turkey" and Ayşe Öncü, "The Politics of Urban Land Market in Turkey: 1950–1980," *International Journal of Urban and Regional Research*, 12 (1988), 38–64.

16. Buğra, "The Immoral Economy of Housing in Turkey."

17. *Kadın* March 15 (1947): 12.

18. Uğur Tanyeli, "Osmanlı Barınma Kültüründe Batılılaşma-Modernleşme: Yeni Bir Simgeler Dizisinin Oluşumu" in Yildiz Sey, ed., *Housing and Settlement in Anatolia in Historical Perspective* (Istanbul: History Foundation Publications, 1996), 284–297.

19. State Institute of Statistics, *Statistical Yearbook 1942* (Ankara), 115, 458.

20. Anonymous, *Kadın*, April 25, 1959.

21. The internal report prepared by Arçelik on "The Evolution of the Sales of Household Durables and Market Shares," Istanbul, May 1996.

22. The price of the first domestically produced Arçelik refrigerators was 4,425 Turkish Liras (U. Ekşioğlu, "Arçelik'te 27 Yıl," *Ayda Bir*, October 1985). This was considerably higher than the annual wage earnings of a coal miner (TL 3,906.6) and only slightly lower than those of a textile worker (TL 4,831.2). The average annual wage earnings that include the considerably higher ones in commerce and finance, were of the order of TL 5,198.4 (calculated on the basis of daily earnings given by The State Institute of Statistics, *Statistical Yearbook 1963*, Ankara).

23. In 1965, e.g., 24.57% of households in city and district centers did not have electricity and 42.31% did not have running water. Naturally, these percentages would be much higher for households in rural areas. The State Institute of Statistics, *Statistical Yearbook 1965* (Ankara), 48–49.

24. Rosa-Maria Gelpi et François Julien-Labruyère, *Histoire du crédit a la consommation: Doctrines et pratiques* (Paris: Découverte, 1994).

25. The following discussion is based on personal interviews with managers and dealers of Arçelik as well as on internal company documents. Ayşe Buğra's, "Non-Market Mechanisms of Market Formation: The Development of the Consumer Durables Industry in Turkey," *New Perspectives on Turkey*, 19 (1998): 29–52, also uses the same data set.

26. For example, the special issues on the annual meetings held in 1984 and 1985 of the company journal *Ayda Bir*, February 1985 and December 1985.

27. Pulat, *Dar Gelirlilerin Konut Sorunu ve Soruna Mekansal Çözüm Arayışları*, 282.

28. Buğra, "The Immoral Economy of Housing in Turkey"; Öncu, "The Politics of Urban Land Market in Turkey" and Tansı Şenyapılı, "Örgütlenemeyen Nüfusa Örgütlü Çözüm: Çözümsüzlük," *Konut Araştırmaları Sempozyumu*, Konut Araştırmaları Dizisi 1 (Ankara: T.C. Toplu Konut İdaresi Başkanlığı, 1995).

29. Buğra, "Non-Market Mechanisms of Market Formation."

CHAPTER 6

Home in the 1990s Galilee: An Ethnographic Approach to the Study of Power Relations

Tania Forte

T his chapter starts from the premise that people, in the Middle East as elsewhere, are historically constituted; it shows how this is made evident in ways people fashion relationships to their home.

Foucault's studies of the relation between people and institutions have caused us to ask how we are shaped through internalized disciplines and desires, manifested in everyday life, as well as through multiple sociopolitical power struggles. To what extent can these theoretical insights apply to the discourse on homes and houses in the context of late twentieth-century states? I approach this question by an ethnographic exploration of the relation between Israeli state institutions and the everyday life of Palestinian citizens as it is manifested through the production of homes. The constitution of homes provides data for an analysis that encompasses changing conceptions of sociopolitical categories, households, and individuals, and which also reveals how these are shaped by the application of state power through land control, planning, and mortgage financing. The data I use here was collected mainly during ethnographic field research in the Galilee in 1990–1991 and 1999–2000. I examine here the constantly changing activities that enter into the production of homes as "the ensemble of more or less regulated, more or less deliberate, more or less finalized ways of doing things, through which can be seen what was constituted as real for those who sought to think it and manage it and the way in which the latter constituted themselves as subjects capable of knowing, analyzing, and ultimately altering reality."[1] Thus I analyze "ways of doing" through which people make homes in relation to "ways of thinking" about what it means to be an acting subject in the state of Israel at the end of the twentieth century. I pay close attention to daily practices, through which I elicit conditions under which people in the Galilee come to imagine and shape selves and social relations on the one hand; and on the other, I trace transformations in power relations, through which people come to experience themselves and others differently over time. What such an analysis can provide is a better understanding of power relationships involved in everyday life practices by which people make themselves as they also build the homes in which they live. I gathered most of the ethnographic data on which this article is based while living in the home of Sayideh and Hasan Othman from the town of Deir al-Asad in the Galilee.[2] In the following pages, I use their housebuilding experiences, and those of their extended family—as I witnessed and understood them—as an instance of the processes I am tracing here. I frame these experiences within historical, political, and socio-economic contexts that give a better sense for what these experiences tell us about the making of selves in the state of Israel at the turn of the twenty-first century.

Because most of my analysis is based on ethnographic data of the everyday practices through which homes are made and maintained, I draw on recent discussions of person–object relations in anthropology and related disciplines. First, homes can be studied as important objects of sociocultural production that provide opportunities for people to reassess selves and social relations, and attempt to influence their definition. Studies of such objects have been extant among anthropologists concerned with consumption,[3] which tend to stress the creative bricolage through which consumers give objects cultural meaning. The endless discussions about the making of homes in the Galilee certainly seems to demand such an analysis.

But houses, as much as they are objects of consumption and desire, are nevertheless also objects deriving from various visions of political power that have little to do with individual choice. Thus many contemporary states apply power directly through policies of control of space, which include zoning and taxation, building regulations, social programs of home financing or public housing. The practices by which such power is applied, and the various sociopolitical hierarchies and reciprocity relations through which it is made manifest, should also be taken into account in our study of homes (see also Buğra, chapter 5, this volume).

Therefore I posit here that homes themselves are the product of power relations that are constantly reassessed and manifested in the course of minute decisions made about money and its meanings, about people and their roles in the family, and the building process. Historians working on consumption have recently addressed issues of power to show how people negotiate selves in relation to macro-level political and economic trends in the context of everyday life.[4] Such studies differ from structural studies of houses that are classics in anthropology,[5] and more recent ethnographic studies of homes,[6] precisely because they examine sociopolitical power relations at a given time, and trace historical change in these relations as they are manifested in new desires and their everyday practices.

This chapter develops four approaches to understanding the workings of power through house-building in a Palestinian town in the Galilee. The first considers how a genealogical notion of "house" has changed over three generations, and traces how this notion relates to sociopolitical power relations inside and outside the town. The second explores how sociopolitical power relations are manifested through control of space, and through economic and social assessments of the value of land and building. The third part traces how state and global economic trends create desires for certain homes, and how in turn these desires reframe power relations between family members. The final part presents ways in which homes constitute specific inscription of these desires, and traces how they operate to shape sociopolitical distinctions.

Deir al-Asad is situated in the Western Galilee, and is surrounded by a relatively large Jewish town founded in the 1960s, Karmiel, several Palestinian towns, and a few small Jewish communities started in the 1970s on neighboring heights and known as *mitzpim* (observation points). At the time of my research Deir al-Asad had some 7,000 inhabitants, who were categorized in the state census as Israeli, Arab, and Muslim—that is, their minority status in the state of Israel is conceived in terms of ethnicity and religion. Like most other Palestinian towns in Israel, Deir al-Asad is characterized by low geographical mobility, since discrimination in Israeli society makes it especially difficult for Palestinian citizens to find housing and develop support networks in mostly Jewish towns.[7]

Although the income of Palestinian citizens of Israel has increased over time, it still lags behind that of the population defined as Jewish, and a large proportion of them continue to work in low-paying jobs in the service economy, industry, or agriculture. In addition, some—though a smaller proportion than among the Jewish population—have been able to establish businesses or enter middle-class professions. Because Palestinian citizens are often perceived as a security risk, they are more likely to be entrepreneurs than professionals, and to work in liberal professions than in other fields.

Since the 1990s, an increasing number of women were working as well, mostly as teachers or unskilled laborers, supplementing family incomes. As I explain later, changing desires and social norms have affected the way people conceive of proper homes. Such notions are manifested in changing notions of social groups as well.

Tensions of Power: Geopolitical and Social Cohesion

Deir al-Asad, 2000. Sayideh and Hasan are going to build. They have been married for twenty-four years, and as first cousins they grew up close together. Their parents built houses in the 1950s, on a plot of land that once belonged to their common great-grandfather, Othman. This plot is where Othman's son Rashid first built his home sometime in the early twentieth century, right outside the walls of what was then a growing town, between the mosque and the ruins of the monastery/fortress. Rashid's plot is now a street. The men, and many of the women who live on this street, are first and second cousins, descendants of Othman. In fact, Othman is the family name that became inscribed on their identity cards during the British mandate, and they are known in Deir al-Asad as Dar Othman—literally the house of Othman.

Today in Deir al Asad, buying land and building a house almost always involves the negotiation of social relationships within the extended family, as well as more formalized exchanges of work with other people having specific expertise. It involves the ability to secure available land, and a certain amount of money for construction. It involves state power, such as the will of government planning authorities who decide how much land is to be released for building purposes, and where that land will be located within the town. One's home is seen as a concrete representation of one's social importance within Deir al-Asad, a part of which is the ability to express one's individual tastes and styles, and another part of which is one's ability to create a *dar*—an extended household of one's own. Personal tastes and desires, then, do not operate alone here. Rather, they interact with social networks within which relationships are produced and invoked, and where economic and social exchanges are practiced. The sociopolitical influence of a *dar* is related to the purported power of its members, which is also manifested and enhanced by the concrete house on the landscape.

When Hasan and his older brother Mohammad got married in 1969 and 1970, their father built an extra floor in his home, with two rooms. Each couple took a room, and they shared a balcony. The bathroom and kitchen facilities were shared with Muhammad and Hasan's parents.

Deir al-Asad is, like other Palestinian towns in the Galilee, patrilocal. When women marry, they leave their parents' home and become a part of their husband's family. But what it means to move in, and what it means to "open a house," as the expression goes, has much changed over time. Consider, to start with, the place, the time, and the meaning of building.

Hasan and Muhammad's grandfather Rashid built his home on his father's land at the turn of the century. At the time, the house was surrounded by a large courtyard and some olive trees. When his children married, extra rooms were added, and the household shared work, tools, and provisions. Rashid's sons had lived with him, and only built their own homes in his old age, when their children became of age and needed rooms of their own.

The neighborhood in which Sayideh and Hasan currently live is that of their extended family. To them it is *al-hara*, the neighborhood. Outsiders refer to it as Dar Othman (the house of Othman).[8] Within that extended family—which the street or neighborhood indexes—are several house buildings. These houses are a concrete manifestation of Dar Othman, much transformed since the 1950s, as the communal courtyards have disappeared and new generations have built ever-larger homes. But these individual homes are also called *dar*, for example when Sayideh prosaically asks her son to go and get some sugar from *dar* Muhammad upstairs. Each is, potentially, an extended family or *dar* in the making. Both "house building" and "households" (*dar*) are important, dynamic sociocultural concepts that have been reproduced and transformed over time. Such concepts are important because they represent the socioeconomic and political power of the Othman family as it was manifested at the turn of the twenty-first century in Deir al-Asad. Such power, as I explain later, is both shaped and constrained by institutions of the state of Israel.

With one exception, the concept of *dar* has not been discussed in social scientific literature on Palestinian towns. Scott Atran defines *dar* as an extended family that reaches back two or three generations to a common ancestor.[9] He remarks that this concept is more often used locally than the *hamula* [or clan] described by travelers at the turn of the century, and subsequently used as a political unit of analysis by anthropologists working in Palestinian towns in the 1950s or 1960s, and further critiqued by Talal Asad in the early 1970s. *Dar*, argues Atran, doesn't appear to have had much importance on the political level, nor has it attracted the attention of social scientists concerned with the political structure of Palestinian settlements.

One reason *dar* has not attracted much attention is because the concept itself is more dynamic than Atran's definition leads us to believe, and therefore harder to pin down. It appears to have multiple meanings on the one hand, and to have changed over time on the other. For instance, *dar* can refer to a "nuclear" household, or in the case of Dar Othman to an extended family going back five generations. But in 1990s Deir al-Asad it is often used as well to describe the largest possible groups in the town, Dar Asadi and Dar al-Dabbah, each encompassing about 3,000 people in 2000. One word I rarely heard in this context was *hamula*, a word Israeli journalists and some social scientists still typically use to describe the extended family. In contrast, *dar* was part of practically every casual conversation. How then can we unpack the meaning of such a term?

Considering the *hara*, the neighborhood where Sayideh and Hasan live, offers a possible, if partial, explanation. At the time of Rashid, there was one house on the plot. At the turn of the century, in Rashid's days, there were perhaps 350 inhabitants, and Rashid, whose household was already known as Dar Othman, was part of an "extended family" known at the time (and occasionally still known today) as Har al-Sharkiyeh, the eastern neighborhood. It is represented as a large subgroup of Dar Asadi on a genealogical tree dating from the turn of the century. In 1948 there were 1,100 inhabitants.[10] At that time Rashid's three surviving sons lived there with eleven grandsons. In 1991 there were over 6,500 townspeople, and Rashid's living male descendants numbered sixty-eight. The *dar*, then, is about the same size as Har al-Sharkiyeh must have been at the turn of the century. It is materialized on the landscape as a street instead of a house. *Dar*, thus, is not a static concept, simply reproduced as extended families break up. It is a concept that has been both reproduced and transformed through the generations.[11] Today, Dar Othman operates as a social and political unit within the town; Har al-Sharkiyeh, now too large, does not.

The dynamic cohesion of groupings within the town is visible on the ground. Thus the upper neighborhood was often referred to as the home of Dar al-Dabbah—the place where two Dabbah brothers had settled in the eighteenth century. Likewise, until very recently, people from Dar Othman tried hard to build homes close to each other. Othman had bought the piece of land on which his son built from Dar al-Dabbah, and his children and grandchildren also made their homes there. But his great grandchildren didn't have enough land, in the 1960s and 1970s, to build the new concrete block houses that had become the townspeople's idea of a home then. Sayideh's brothers exchanged a piece of building land they owned in the upper part of the town with one adjacent to Othman's plot. Their cousins bought another adjacent piece of land, at around the same time, from someone in the town as well. At that time, even as sons built houses and made money of their own, they continued to share many family resources: space, childcare, food preparation, chores, and most often finances. It made sense that the cousins would chose to build close together, and were ready to exchange or buy land in order to do so. Their actions transformed the place in order to match the relationships and practices that made up the extended family at that time. Yet because according to the state building code buildings in Deir al-Asad could not exceed four floors, the *hara* could not accommodate the current generation. What becomes evident is that social cohesion is not "given," nor is the social map of a neighborhood simply an outgrowth of "natural" relationships: generations have shaped these relations with great effort as land has become scarcer and as government institutions have constrained them.

This, then, leads to an observation that flies in the face of analyses that consider the "clan" as a traditional, continuous category of Palestinian society. The fact that *dar*, conceived as an extended family group, functions as a political and geographical entity within Deir al-Asad is not only an outcome of population growth. As Talal Asad argued, the continued existence of Palestinian clans as political units is also the result of the activities of Israeli political institutions within Palestinian settlements in Israel.[12] While Asad's work referred to the elections apparatus, mine extends this conclusion to the relation between state land policies and local social relations. One reason why people keep living in the same place, strengthening social cohesion, is that they cannot easily move elsewhere within the country. Thus the geopolitical discourse of the state structures the feelings and practices by which people organize social cohesion in town. Social cohesion itself, then, is dynamic and historically constituted by power relations between the state and citizens in the town. In order to show how this works in practice, I describe in the following the impact of state policies on the ways in which townspeople perceive and practice their ability to build homes.

State Control and the Capacity to Make Homes

Long-standing national political tensions determine the movement of land and houses in Deir al-Asad. In an effort to make nation and territory coincide, state land in Israel was conceived as the "inalienable property of the Jewish people," held in trust by state institutions who were not authorized to sell it.[13] As a result, land officials adopted a British leasehold model under which homeowners obtain forty-nine or ninety-nine year leases rather than property title to the land.[14] This was intended to provide continuity between generations without ultimately "alienating" land from its purported owner, the Jewish nation. Officials consider land owned before the state of Israel to have a specific, "private" land status; all other land—including land appropriated by the state over time—is seen as belonging to the state exclusively.

On the surface, land in Palestinian settlements in the Galilee is regulated by the same laws as land in Jewish settlements. Yet government land policies have been applied very differently. For instance, planning of Arab towns has been notoriously constricted, creating serious space problems for private as well as public facilities.[15]

Hasan and Muhammad are the first of five brothers. After they built their own home in 1970, the next two brothers married and moved into the small apartments over their parents' home that Hasan and Muhammad had vacated. They expected, in turn, to save and build their own building on the remnant of their fathers' courtyard. Yet that transition was difficult: this space had been designated as a "green area" by the municipality. This was because the area was close to the Khan, the ruins of the old crusader fortress upon which Deir al-Asad was founded, and was protected as an archeological site. Since the family didn't have sufficient political clout with the local council, they were not able to prevent the expropriation. In exchange, they received a forty-nine-year lease on a plot further up the hill, on land that had belonged to the mukhtar, the town leader, before 1948, but thereafter belonged to the state as "absentee property." Ultimately, because public land is virtually nonexistent in Deir al-Asad, the town leaders obtained permission to build a Senior Day Center and a mother and child clinic on the brothers' plot.

Public space is at a premium in Deir al-Asad, and very little land has been designated for public use. State land here is mostly land recuperated from people categorized as absentees. Some of the public buildings in the commercial zone were built on such land. But since most land there is private, businesses and institutions—including the local council—often occupy the bottom floor of a private building and pay rent to its owner. Schools were also built on "absentee" land, but in the early 1990s there were not enough classrooms to accommodate the number of students and the local council rented private living rooms in which classes were held. Through differential budgets, as well as through differential land allocations, government officials have constricted the expansion of Palestinian settlements; this constriction, in turn, manifests itself in everyday lives.

One main tension in securing land for building involves government land control policies on the one hand, and local demographic expansion on the other. Because of this tension, people live in increasingly crowded conditions. On the one hand, people wanted and were expected to produce larger homes of "their own," and since the mid-1950s found the means—financial and otherwise—to bring them into being. But in addition, the town population grew tremendously. The price of land greatly increased. And, most importantly, the government restricted building zones in most Arab towns in such a way that the area available could not accommodate the townspeople's present numbers and aspirations. The nearby town of Karmiel has been regularly expanded as new neighborhoods were built to attract [Jewish] inhabitants. In contrast, according to the district engineer, most areas around the current Deir al-Asad and Bei'neh building zone were slated to be kept free of buildings as "green areas," possibly to be reforested.

The Israel land authority has often handled expropriation for state and municipal needs through land exchanges. Yet Palestinian Israelis complain that on the whole state officials heavily load the terms of land exchanges, since they give up what they consider to be "full" property titles for forty-nine-year leases. In addition to exchanges in case of expropriation, Israel land authority officials have conducted "exchanges" with townspeople who need building land inside the building zone: they receive plots of half dunam of "state land" on forty-nine-year leases if they give up a larger quantity of land they own fully, but which is zoned "agricultural." The exchange rates are purported to reflect the rising prices of land in the towns as the building zones have not grown as fast as the needs of the town people warrant. Yet again, state control determines the terms of the exchange: a variety of [Jewish] state officials determine both the zoning and the rate of exchange, and Palestinians strongly criticize these practices of control.[16] The relationship between power and exchange practices is evident here. As the terms are set by state officials, townspeople only exchange when they have no other choice. Even though exchanges are discussed in terms of equivalent pieces of land, and there is an official contract to reflect this fact, the practice itself belies this. Here, as in the many practical historical and geographical contexts in which officials or authorities have determined measures,[17] the stronger party can easily rationalize exchange and assessments in a way that is advantageous to controlling elites. Officials have the power to determine the measuring unit, the need for exchange, and its terms. By extension, such power laden terms also work to structure the feelings and assessments of townspeople in relation to the land on which they stand to build their homes.

People's sense of discrimination and insecurity due to land control practices in the town is evident in their own ways of valuing land. Until the mid-1990s, the price of land with an ownership title dated before 1948 was significantly higher than that of land leased from the government on a forty-nine-year lease. This reflected in part a desire to preserve what was seen as a more "legitimate" pre-state ownership embedded in local social power relations, and in part a distrust of the greater government control exercised through leases, which were perceived to provide less stability of ownership since they could be terminated. It also reflected a deeper historical tension, the result of a previous struggle over land won through war and conquest: most building land leased by the government within the town had once belonged to "absentees," who, several older people told me, might at some point claim it back. By 1999, land had become so scarce and expensive that people barely made these distinctions—where no land was available, any land would do.

These days, it is very hard to buy land from private owners in Deir al-Asad, mostly because, in the state-designated building zone, there is little such land to be had. People hold on to whatever land they own as much as they can, since, as they see it, it represents their children's future. In 1998 the price reached $150,000 a dunam, more than what building land costs in the nearby town of Karmiel and three times what it cost ten years before. In the past fifteen years, people in Deir al-Asad and other towns in the area have been talking about building tall apartment buildings, but so far no funding has been made available, and none have been built. People who don't have building land are either tearing down old houses and building larger, three-story buildings on the same plot, or using makeshift additions. Yet for many young men their ability to marry depends on their ability to build houses. So why don't people move elsewhere?

People from the area, and especially from the nearby town of Majd al-Krum where land in 1999 sold for $300,000 a dunam (if it is available for sale at all), have been thinking about this. Many mention that the hard part is leaving behind the major network, one's family. But those who feel they don't have a choice have started to buy land to build in other towns and especially in Sha`ab, a nearby Palestinian town where the going rate is less, and where the state is leasing land on a long-term basis.[18]

The political tension at work here shapes the movement of homes and people on Deir al-Asad land. Whether or not they are aware of this, most state officials couch control within a scientific discourse of planning and land rationalization that masks the fact that they impose discriminatory practices. Thus in the late 1990s several Jewish engineers deplored the few planted areas still visible in the town building zone and advocated that the local landowners should sell these plots rather than holding on to it for their children. Further areas could only be open for expansion, they argued, when such "wasted" space was filled. Town landowners, on the other hand, were concerned with safeguarding future building space in a context in which it is made both scarce and expensive.[19] Thus the tension between state land control and Palestinian demographics stood out as a pressing and explicit concern. A third tension, in turn, was that expressed between a recently built neighborhood, perceived as "modern," and the old family neighborhoods, perceived as "traditional."

Of the households from Dar Othman of Sayideh and Hasan's generation, four have so far managed to get land and build elsewhere, and an additional three—including Hasan—are planning to build on the same newly acquired piece of land, much further up the hill, near the new school. In the 1990s, six houses in the hara have been expanded to enlarge the living space and to build extra apartments for the newly married children. All the other households, it is expected, are saving and working in order to buy land and build. As they do, the houses they leave behind will be remodeled for younger brothers or nephews who are coming of age.

Last year, Sayideh and Hasan's oldest son Mahmoud, who is working as an electrician and finishing his computer engineering degree at the Technion, got engaged to his mother's brother's daughter, who lives four houses down the street and is studying to become a kindergarten teacher. He entered the lottery for plots of state land designated for young couples.

It was rumored that that land would be sold to the winners at about 35 percent below the current market rate. This is part of a new program that the government is supposed to be testing here. One lottery for sixty plots was held in 1996, but no one is sure when the next one will be, and the list of registrants is getting longer. In 1996 the odds were one in three; in 1999, according to townspeople who repeatedly and anxiously check the numbers of candidates registered with the local council, the odds were about one in ten or twelve. As of 2001, the lottery had not yet been held.

As building land in the center has become scarce, most new house builders are moving out to the only newly designated building area, further up on the hill. Building away from the family is not always considered a hardship. In fact, some people—mostly younger, but also some middle-aged—claim that they prefer having neighbors who are not extended family as they will not get involved in the family's business. The new areas are considered "modern," with all brand new houses.[20] To these are attributed all the "advantages" of what people perceive as Western individualism. They boast that they can close the door, have friends over to visit or go visit others without having to account to relatives; as less information is directly available to members of the extended family, there is less gossip and fewer arguments. While they express some preference for living in the new area away from the extended family, new couples would still, for the most part, not consider living in a different town. Some closeness to a family network and its many—if often unspoken—advantages are still relevant to them. Thus the purported advantages of distance from relatives, often described as freedom, modernity, cleanliness, announce an emergent desire for greater privacy. Yet the same people often choose to participate in the "old neighborhood" life as well: at times they bring their children for babysitting, borrow cars, use common spaces in the *hara* for celebrations, share bank accounts or distribute foodstuffs, and otherwise continue to participate in extended households' exchange networks.

Houses are thus the product of tensions in changing social and political practices: on the one hand tensions between emergent personal desires and emergent normative practices, on the other tensions between demographic expansion and repressive government land control practices, and finally tensions between family network neighborhoods seen as "traditional" and the newly created nonfamily one, viewed as "modern." All of these together shape the movement of houses on the landscape.

Financing the "Modern" House: Money, Household Labor, and Meaning

Sayideh remembers how tight their building budget was, even after she sold the gold from her wedding. On days when she took her infant son Mahmoud into town to see the doctor, she walked the five miles because she didn't have bus fare. At that time, she started the sewing workshop she has maintained ever since on the ground floor of their building, and made extra money to pay for the household expenses.

In 1972 Muhammad and Hasan's households paid for the foundations and the shell of the three-story structure. Then each household "closed" its own apartment in time, paying for these expenses separately. Neither household sought the low-interest mortgages available through the housing ministry. At the time, remembers Sayideh, people were afraid to borrow from Jewish banks, a fear to a great extent connected to the experience of the entanglement of state control in their personal and political lives.

The money used to finance house-building is not experienced as neutral quantities that can be added together. Instead, I show here, it is linked to activities or practices out of which persons and sociopolitical worlds are constituted. In her book exploring social practices through which money is valued and circulated in the United States, sociologist Viviana Zelizer traces ways in which money is "marked" by social and institutional networks.21 In Deir Al-Asad, money is marked by social relations, but also by political ones. Tracing these relations can then help us understand how power relations enter into the constitution of people, and how people reconfigure them as they can.

Money from the state, for example, reflects particular ways in which the state has constructed its Palestinian citizens and in turn the ways they have responded to practices connected to this categorization. Between the 1950s and the 1970s, state-subsidized low-interest mortgages were available to all Israeli citizens. But then as in the 1990s, the amount of the mortgage was systematically less for most Palestinian citizens of Israel. Higher sums were available for new immigrants, for Jewish religious seminary students, and for people who had served in the Israeli armed forces, or whose relatives had served. This effectively included all the Jewish population, and excluded a large majority of the Palestinian citizens of Israel. Most townspeople are aware of such differential treatment, but as far as building houses goes they accept it as part of the reality in which they live.

Government control practices included surveillance of townspeople's lives through local informers. In addition, the legitimacy of their claims to land ownership was undermined as much of the townspeople's land was expropriated, and their entitlement to land systematically questioned through the courts in the 1950s, with some court cases lingering into the 1980s. Unlike Jewish immigrants, Palestinian citizens were not encouraged to use government mortgages. Ordinary people who didn't have privileged connections with government or political figures saw mortgages or bank loans in general as strings attaching them to the state in dependency relationships, which in turn made them vulnerable because they could be used as leverage against them. Most townspeople started to use government mortgages in the 1980s. By that time government surveillance had become less aggressive and people felt more confident in establishing their civil rights. By then, government moneys were not widely considered to endanger household status. No longer perceived as negative value, they were seen as enhancing the capacity to establish one's home. These perceptions, it seems, were directly linked to the townspeople's experience of the conjunction between various government practices toward them—on the one hand, the use of mortgages, on the other, the political control with which it was associated at one time, but not at another.

Another string attached to home financing is a woman's jewelry. Sayideh's gold, which she contributed to the house, is another money contribution that entails sociopolitical meanings and connections. According to local ideology, the jewelry was given to her by her husband's household and she can dispose it as she sees fit after the wedding. Jewelry is sometimes given with the wish that a young woman may never "need" to sell it. At the turn of the century a woman's necklaces, bracelets, headdresses, and part of her wedding costume, often consisted of gold and silver coins that could be displayed, but also stringed together or pulled apart when needed later on.[22] They were both money and adornment.[23]

The same ambiguous meanings are present in women's jewelry today. On the one hand, it reflects the ability to provide for the family she is entering— it is an object of substantial value displayed together with her. On the other hand, it is money, since it can be sold and converted into other things—thus women in the 1970s often sold some of their jewelry in order to build a home. On the one hand, it is explicitly presented as hers; on the other, it can be put to the use of her husband's household if the need arises. So the gift, in fact, is not necessarily made to the woman herself, as an individual, as is often portrayed both by local people and by anthropologists and historians. It becomes meaningful through a woman's actions, in this case as she converts it into a contribution to the house for her new household. But why would a woman contribute something that is seen as her property to something that is seen as the property of her household? What is most important for her and for the household, as I will show later, is not the gold itself, but the way she uses it and the relations it creates.[24]

Sayideh sold her jewelry in the 1970s. In the early 1990s, one of her husband's brothers was criticized because he kept buying gold for his wife to show how much he loved her, but as a result he could not afford to build a house (see the second section earlier). In the late 1990s, at a time when her household's financial situation was good, Sayideh who had earlier asserted that she didn't care about gold started to explain that a man who bought gold for his wife was "romantic." Since her husband had not taken the initiative, she then pressured him to go and buy her several heavy gold bracelets. At the time, she justified this by recalling that she had sold her jewelry in the first place for the family, and it seemed only right that she should now be "given" gold again. In her tellings, she made the jewelry mean different things: when she sold it, she was doing what was morally and socially proper for the household, unlike the relatives who were criticized for buying jewelry instead of building a house. When she made her husband buy her gold, she elicited it as an expression of love and romance. The gold she received then was not simply a replacement for the one she had sold, though indeed it did have an important financial value. It was a "romantic" gesture, an expression of his attachment to her, which returned the attachment she had shown for the household when she had sold her jewelry. It was, then, an action replacing her own action twenty years previously.[25] The money she contributed in the 1970s was a social action rather than an object or a quantity. It was not a simple debt that was being repaid, but rather a person and a relationship that were being recognized and enhanced. Thus sociopolitical actions—be they state practices or a woman selling her jewelry—enter into the financial production of the house. Other actions by which the house is financed include extended family relations. What is crucial here is the dynamic redefinition of persons that my exploration of state and social practices, based on the dynamics of power, allow me to trace here.

As far as getting hold of building land goes, Sayideh and Hasan were lucky. In mid-1998, before the land lottery took place, they were able to buy a small plot of land to build on. It happened like this: Sayideh's brother Qassem was doing good business into the mid-1990s and had bought a plot of 1.4 dunam, on which he intended to build his new home. Business difficulties compelled him to sell at the end of 1997, and his brother Ahmad bought the land for the going rate. Ahmad's meat factory was booming then, and he had just built himself an over- $300,000 home on one dunam of land he had purchased from another family in the early 1990s. He bought Qassem's land, he says, so that it would not go out of the family. In 1998, he agreed to sell it as follows: 0.4 dunam to Sayideh and Hasan (his sister and cousin), and 0.5 dunam each to his brothers Qassem and Adib. Sayideh and Hasan bought their piece for the equivalent of $120,000 a dunam, slightly below the market rate (because "Ahmad is our relative, and we help each other"), which they agreed to pay in full within the year (because "Ahmad has large expenses on his own house"). Adib bought his piece for the equivalent of $80,000 a dunam. His discount is larger, Sayideh explains, because he is Ahmad's brother rather than his cousin, and he works for Ahmad as well.

Social connections and their meaning are very much present as money and land change hands in the extended household. While Ahmad and others mention causes for the exchange, another unspoken cause for this selling at less than the going rate is plain. Ahmad's gesture preserves, and even enhances, the extended household's dignity. He gets his brother Qassem out of financial difficulty and enables him to build a home, albeit a smaller one. He helps his sister and cousin to build. Sayideh underlines that it is because her husband is a cousin (father's brother's son) that they get the discount from Ahmad, rather than because she is his sister. Indeed, as a sister she holds an ambiguous status here. On the one hand, she should not expect preferential treatment when it comes to property division since she is not part of the exchanges usually expected between brothers. Under normative circumstances she would have been expected to share a plot with her husband's brothers rather than with her own. On the other hand, Sayideh is especially close to her brothers, and very much involved in "helping out" in myriad informal ways. Her own actions and personality have a great deal to do with her brothers' agreement to have Sayideh and Hasan buy into the plot. Yet she must be very delicate about the formulation of this transaction. If she says that they received the plot because of her, her husband might be branded as "weak." It is therefore important to explain the deal in terms of Hasan's relationship to her brother rather than in terms of hers.

Since she is a wage earner and a strong woman herself, Sayideh has become particularly aware of the problem of representing her household through such statements. When she first started earning money in her tailoring shop, she boasted to relatives that her earnings allowed them to buy a new refrigerator, a freezer, and a washing machine. Rumors then flew around the neighborhood that Hasan was not able to provide adequately for his family, and as a result he insisted that she close down the workshop. When she started to work again, she suggested that she would use her earnings for routine expenses such as food or utilities, and that he would use his to purchase new furniture or appliances. In the same way here, she presents the discount Ahmad gives her household as the result of his connection to Hasan rather than to her. The fact that the discount is less than it would be had Hasan been Ahmad's brother supports this assertion and discourages rumors as well. Here it is not simply connection that is expressed through the financial arrangement, but a socially proper connection. Highlighting this connection effectively downplays Sayideh's contribution to the procurement of land in one case, and to the purchase of appliances in the other. Here the meaning of her actions is hidden so that the status of the nuclear and extended household can be enhanced in the normative terms of local discourse. It is possible, though not certain, that this discourse may at some point be stretched to include Sayideh's now somewhat eccentric—though obviously socially powerful—actions. Meanwhile, they are hidden.

In the same fashion, Sayideh and Hasan can say that they are helping Ahmad make payments on his home. And best of all, no one in fact had to sell the land on a public market: it remained "in the family." In this way, Ahmad can enhance his status as a wealthy, wise, and generous person within his own family as well as within the community.

So Sayideh is taking in extra work. Fancy curtains have become fashionable and she is doing brisk business in them. In addition, "bridal parties" with bridesmaids have become the norm at the end of the 1990s, and Sayideh is quite busy in the summer making sets of bridesmaids' dresses. Hasan found a second job on some nights and weekends, and Shadi and Fadi, the twenty-year-old twins who dropped out of college the previous year, got jobs in a factory and in construction. In order to pay for the building itself, the three older boys and the parents will take mortgages from the housing ministry.

Though discounts, mortgages, and savings may help to finance the house, producing the house rests mostly on their own labor and that of their extended family. In the Jewish areas of 1990s Israel, construction was done either by developers or by contractors engaged by the future homeowners. Whole neighborhoods, or even towns, were developed and sold to the [over-whelmingly Jewish] public. Those who want to build their own home do so by hiring one or several contractors, the cost of which was estimated at $800–1,000 a square meter in the 1990s. Indeed, construction has been a major sector of employment for Israeli Palestinians. For the amount of money it would take to finish one 140 square meter apartment in a Jewish area, Sayideh and Hasan will build two or three apartments. Two apartments, Mahmoud's and their own, will be finished right away. The others will be completed when the twins marry. Now the twins "help" their older brother; later, he will "help" them.

Unlike most Jewish Israelis who are building their own homes, Sayideh and Hasan are not calculating cost in square meters nor working through one contractor. Certainly, the size of apartments is constantly talked about and compared in square meters. But when discussing costs people in Deir al-Asad add up the cost of making the foundations, of the materials for the building shell, and of other materials they will use in the coming year.

Hasan can get cheaper iron from his job. Sayideh's brother Said, who operates a quarry and deals in *balat* (floor tiles) and *sheish* (the Hebrew word for marble) can help provide cheaper floors. Their son Mahmoud, who got an electrician's license before going to the university, will purchase materials wholesale and lay down the electrical system. Since he has helped others to install electricity in their new homes, people skilled in plumbing and masonry can be expected to return the favor. The household will hire workers, especially for digging foundations and carting away dirt, but will also work on the house. As someone commented, there isn't one Darawi (inhabitant of Deir al-Asad) who has not worked in construction.

Of course, Hasan and his sons will work on the building, and cousins will help as well. Many indeed have worked in construction, and most young men have experience helping out on relatives' house-building projects. This kind of work is informal, sporadic, irregular, fluid: when relatives help out, sometimes they are paid, especially if the relatives who are building are more well off than their own family and can afford to hire them fulltime. But most often they work with the unstated understanding that the relatives will help when they in turn have to build. In 1999 on a cool spring day, for example, I saw one of Sayideh's brothers, in his late forties, working on his nephews' new home. He himself has taken almost fifteen years so far to build his over $400,000 home, which will house his family and those of his two sons. The nephews have been helping. While he has paid some of them, he also helps them work on their building.

First Sayideh and Hasan thought they would build quickly, so that Mahmoud could marry and move in. But then, calculating wedding and building expenses, they decided to wait until after the wedding. Sayideh's brother Ahmed has an extra apartment on the first floor, which his brother Said and family used for eight months in 1998–1999. Said's son is also getting married soon, so Said's household is rebuilding and enlarging his building's shell to provide three apartments for himself and his two sons. He is coordinating this with the completion of the house of his brother Samih, who used to live downstairs from him. Mahmoud's wedding coincides with Said's move back to his renovated home, so that the new couple can move into the vacated apartment. Neither Mahmoud, nor his uncle Said, are expected to pay rent. Said tiled the apartment floor as a favor to his brother Ahmad. Ahmad is doing well financially; the apartment will be used for his son later. Ahmad is praised by relatives, whom he knows will not hesitate to help out if the need arises later on.

The exchange of labor within extended households, while very frequent, cannot be entirely taken for granted nowadays. Sayideh prides herself on the fact that she has taught her sons to help each other. She contrasts this approach, favored by her brothers who have been partners in business, have exchanged money and work when building their homes, and have bought agricultural land in common, with that of her husband's family: "My aunt [her husband's mother] didn't teach them to help each other; Hasan and Muhammad, and the others, all saved separately and didn't share their money; many others in the town don't help their brothers build. They are each poor, all work for other people. My brothers have succeeded in business because they work together, they are stronger."

The word Sayideh uses for her brothers' partnership is *sherkeh*. It means company, partnership, but it is also the word used in the old days for cultivation agreements and goat-raising agreements.[26] It often implies a long-term business relationship in which profits are divided and reinvested. The term, then as now, covers a variety of formal and informal arrangements, to which people attribute different moral values. Here Sayideh chooses to give them a socially positive connotation. Success, as she presents it here, stems from the ability to employ the resources of the extended household to further the ends of "nuclear" households. Indeed, until recently, her brothers put all their income in a common account out of which they paid themselves monthly salaries. Recently, as some of their sons became of marriageable age, their expenses greatly increased and they decided to establish separate accounts.[27] According to Sayideh, tensions between the wives were at the core of this. She explains that "when there are problems like that, that's it, everyone goes by himself, it's better that way" [*lema bsir fi mashakil hek, yallah, kull wahad iruh lehalo, ahsan*]. Clearly here the intention is not to promote individualism, but rather to loosen the financial pressure by dividing up the common bank accounts. It is understood that the brothers are likely to recreate the same arrangement with their own sons once they come of age and need to build a house and get married.

Such an arrangement is not unusual: several sets of brothers or father–son teams in the town bring together their incomes and share expenses for extended periods of time. This is Sayideh and Hasan's current arrangement. Hasan, Mahmoud, and his brothers have their salaries directly deposited in a common bank account, and each of them gets some pocket money. Sayideh cooks for everyone, because buying in bulk and cooking quantities saves time and money. She is happy that Intisar, Mahmoud's wife, who is a student teacher, goes along with this, because "some girls don't like it." Indeed, many young women and men consider that eating at one's parents' house or not having one's own bank account are signs of "backwardness" and dependency. But Sayideh is adroitly promoting Intisar's cooperation by "spoiling" her: Intisar keeps her salary for her own expenses. In addition, the household has recently bought a car out of the common bank account "for Intisar," so she can drive to and fro to work independently (Mahmoud, meanwhile, has the use of a company car through his job). Thus through the "traditional" arrangement of the extended household with a common bank account, Intisar can display the outward signs of a "modern" woman: she has money to spend on herself and leisure activities, and a means to circulate independently.

Common financial and extended household arrangements, explains Sayideh, is what one household in five does in Deir al-Asad at any given time, especially during periods when large expenses are at stake. But she does not see this arrangement as permanent. When they enter the new house, she declares, Mahmoud will be independent/free (*hurr*): "Finished, he lives alone, he can do what he wants." On the other hand, others, like Hasan's brothers, have opted to keep their finances separate. The fact that such different arrangements exist has often been portrayed in social scientific literature and popular consciousness alike in terms of "traditional" versus "modern" household arrangements, or again in terms of the rise of self-interested individualism. It would be easy to see Sayideh and Hasan, and Sayideh's brothers, as simply repeating traditional arrangements.

I argue here that Sayideh's description demands another interpretation. What has happened is that Sayideh and Hasan's family—together with all the other households who have used this approach, albeit temporarily—have chosen to use some patterns resonating with the social and economic relations of "the old days" in order to economize and to re-produce and transform their household. These arrangements do not hold persons in "traditional" relationships, but rather in creative ones that satisfy both present means and present aspirations: Intisar is a case in point. As she and other townspeople describe it, they are not trapped in the webs of custom. Rather, they have come up with a system through which more resources can be harnessed, and thus enhance the power of households to produce and transform themselves within their limited means and circumstances.

To sum up, the diversity of economic arrangements existing in Deir al-Asad does not simply show a movement toward greater "individualism." Rather, it reflects changes in building and housing practices on the ground, which resonate with changes in material and social aspirations as well. While most townspeople at times couch these practices in terms of aspirations to "privacy" or household independence, this very privacy or independence is most often achieved through pooling common resources for a time. The social and political meanings of money that are embedded in these practices show us that such a process is dynamic indeed. But as it changes, paradoxically, it produces people who are just as involved in the reconfigured financial practices of kinship relations, political relations, and global economic movements. The power of these relations continues to be visible in the desires and creative effervescence of social persons-in-the-making in Deir al-Asad at the turn of the millennium.

New Consuming Desires and Distinctions

In Hasan and Sayideh's extended family in the past three generations, brothers have gone from having a room for their family in their father's mud house to living in their own thirty square meter concrete homes, to building ninety square meter apartments, and now to planning 160 to 300 square meters per "nuclear" household. Such increases in household space are consistent with changes in ideas of what constitutes a proper home in Israeli Jewish society. Local people and officials of Israeli institutions alike have couched such changes in terms of progress, development, and economic expansion. In Deir al-Asad at the end of the 1990s discussions of the making of homes tended to replicate the success narrative of consumer society, or to oppose it. Yet the questions they raised for me were more fundamental: how were these new norms constituted, what did changing desires tell us about changing practices of selves, and what could they tell us about changes introduced in power relations by consuming discourses? In this section, I will attempt to answer these questions. In order to do this, I place Sayideh and Hassan's house-building project within additional data collected from the historical and social contexts of Deir al-Asad.

Houses are, first, telling displays of changes in fortune. In the 1920s and 1930s, according to local memory, only four prominent men "built." Building, at that time, meant to bring master builders, workers, and stone from Safad or Haifa, and possibly to command the labor of many subordinate town families as well. The (at the time) high and spacious two-story houses they produced stood in contrast to the mud houses of other townspeople, who depended on mutual help to raise roofs on houses;[28] plastering and whitewashing were tasks shared by extended families.

The door to the *mukhtar's* house, for example, was designed so that he could ride in and out of it on his horse, and would not be seen walking. Mobility and immobility both commanded his distinctive status: outside he rode, surveying the townspeople's work on his property. In his house he sat, while others came to his *maddafeh*, cleaned his house, brought him gifts, and asked his counsel.

Subsequently, two of these four stone houses were expanded as the owners continued to acquire land and extended their compound to include homes for their children. Another man, by contrast, had to sell all his lands, finally dividing the house in which he had grown up into small apartments for his children, who were, in the words of another neighbor, "one on top of the other like sandwiches." Changes in fate are not looked upon with kindness when status is at stake.

One's house today stands as an icon of household accomplishment. Ma'mun, a middle-aged teacher who lived in an apartment in his father's compound and dreamed of building one day, explained it like this: "Their houses point to them [*durhum dallat 'alehum*], the bigger the better." He then playfully summed this up in a poem for my benefit:

> La-min ad-dar al-kbireh?
> La-Ma'mun,
> Al-Amir ibn al-Amireh.

[*Whose is the big house?*
It's Ma'mun's,
The Emir son of the Emira.]

During the mandate period, average people who had lived in mud houses had helped each other build their homes and repair roofs. To many older townspeople who remember growing up in mud houses, the many new concrete homes appearing in the landscape in the late 1950s and 1960s were a sign of wealth and equalization. One of these townspeople who retired after working in construction for thirty years under the Israelis considers himself "wealthy" today, as he told me: "Until 1945, only 6 houses were made of concrete [*beton*]; they had to bring al-Saffadi to build them. Now it is easy to build, the government helps with mortgages, thank God. Everyone can build, me and the rich alike! [*metli metl al-gani!*]."[29]

This man's associations are richly revealing: he conflated the old stone houses with the later ones built of concrete blocks, like his own, and like the many he had spent his adult life building "for the Jews"; and Ali al-Saffadi, a master builder of stone houses famous all over the Galilee during the Mandate[30] was presented as the one and only builder of all the notables' houses (the most notable of which had been built by others, in fact). In his account the privileged few had now become the many, multiplying through wealth accumulated from their work and from government mortgages. To him, the multiplicity of vast concrete houses on the town landscape symbolized the change in status of average townspeople, from "peasants" into proper homeowners. This propriety, bestowed by property itself, is the very subject of desire. The process of building is also an expression of household positioning within the town, and much time and talk is spent on this. As I showed earlier, this has been going on for at least three generations.

Sayideh and Hasan remember vividly how, when they built their current apartment in the early 1970s, they chose the floor tiles, on which they got a good price, how they decided to lay out the entrance room, the living room, the three bedrooms, and the large kitchen, all in what was at the time a spacious ninety square meters.

For that period, this apartment was state of the art. By contrast, a survey of town homes done in 1967 shows that the majority of houses existing at the time had two rooms for an average of seven inhabitants. Some of these are still used by older people today. The same survey shows that out of 375 homes, ninety-eight were built between 1948 and 1961, and an additional 111 between 1961 and 1967.[31] Two hundred and forty-one homes had running water, but only twenty-four had hot water from solar heaters, thirty-two had showers inside the home, fifteen had indoor toilets, and ten used gas for cooking (except for an occasional generator, there was no electricity at the time in the town).[32] In the 1960s, people like Hasan's parents built rooms on the rooftops for their children when they married. It is only in the 1970s that most young couples started to build their own larger apartments, modeled to a large extent on contemporary Israeli homes. Unlike most Israelis, they had to do this without the help of government-initiated and-subsidized housing projects.

During my stays in the town in the 1990s, the bedrooms were only used for sleeping, and the entrance room was where most of the family activities and casual visiting took place. The living room was where the family watched television, where formal visitors were received, and where informal guests slept. It was about twice the size of the entrance room and the bedrooms. Wealthier people's homes had about the same layout, except for the living room, which was between three and four times the size of others', complete with four or five sofas and four or more chairs. This living room was obviously reminiscent of the old time diwan or *maddafeh*, and presumably made to accommodate the large gatherings their owners, as people of power, counsel, connection, and influence, are able to attract.

Sayideh and Hasan describe their new home as modest. For months, if not years, Sayideh has been looking at homes and thinking about what she wants for her own. A newly arrived Russian Jewish immigrant architect drew the blueprint—he has already built several homes in the neighboring towns, homes Sayideh describes as stunning. If she had unlimited money and land, she says, she would build her house on one floor—there would be no need to build up for the children, who would also make separate one-level homes of their own. The living room, she says, would be large enough to have gatherings and parties, bigger than the ninety meters of her present home.

As it is, the building will be four stories high, with one apartment for each of the three older boys. The youngest son who is now ten will live with his parents. Each apartment will be about140 square meters, which is larger than the ninety square meters in which they now live, but about half the size of the mansion built recently by Sayideh's brother Ahmad on one full dunam of land. There will be the same three bedrooms, but one will be a master bedroom with a Jacuzzi, and the others will share another bathroom with a Jacuzzi as well. It's the fashion, Sayideh says, and it will be relaxing for her husband who does heavy manual labor. There will be a guest washroom as well, and four balconies.

She is debating about whether to include a *heder aronot* (Hebrew for walk-in closet), a recent and controversial addition in new Israeli houses and apartments (some argue it "wastes" space compared to the wall to wall closets of the previous era). There will be a *pinat ochel* (Hebrew for dining corner) in the living room, just as in many new apartments. The living room will also be much larger than the current one, with a view of the valley below: in fact, it will take up most of the added space. Sayideh decided against the *mitbah amerikai* (Hebrew for American kitchen, which opens onto the living room): messes are too easily visible, and so is the woman of the house, whose body motions would be observed as she prepares coffee for her guests. Her daughter-in-law is attracted to the gleaming luxury of the *mitbah amerikai*, with its matching wood cabinets, steel appliances, knick-knacks, and fancy halogen burners. Its perceived disadvantages can be easily remedied by the making of a second, smaller and enclosed kitchen space, the *matbah yom yom* (Arabic kitchen for day-to-day use): here mess and smells can be contained and decency preserved. In addition, she justifies the necessity of the *matbah yom yom*: coffee is better prepared in the old-fashioned way, she says, that is on the *matbah yom yom's* gas stove. Sayideh's brother Ahmad's $300,000 home includes a *matbah yom yom* as well as a *salon yom yom* (living room for day-to-day use).

Current social norms in town bring the "modern," "Western" separation of functions within domestic space to new heights. In the fanciest homes, one living room is reserved for the family and informal visitors, the other for distinguished guests—the expectation, of course, is that people with a higher status in the community would receive such guests. It replicates the large space under the control of a local notable to which people from outside the family are drawn: local men from allied families, groups of notables from neighboring towns, officials, important visitors from other parts of the country or from Palestine, international travelers. It is the potential gathering of all of these network connections, once existent in the *maddafeh*, that resonates today in the formal salons of individual mansions. These are also, albeit to a lesser extent, reproduced in the salons of more modest families.

Sayideh who often prides herself on getting good deals, travels to the West Bank town of Jenin, where things are cheaper, for dental work or groceries. At the same time, she is proud of the 5,000 dollars she spent two years ago on an Italian bedroom set. It was the only one of its kind in the store, she explains, and all the visitors have to step into the bedroom to admire it.

Interestingly, bedrooms in both mansions and average apartments are about the same size, and most of the expansion of space as compared to the 1970s' apartments is concentrated in the salon. In displaying her bedroom set, Sayideh can make a mark in a way she could not in the living room.

In her discussions with other women, the bedroom set situates her as someone who can afford to choose the best and distinguishes her as an original trendsetter. This ability to introduce something unique in the town echoes the mansion owners' positioning through style. It is intended both to attract others and to set oneself apart. Not surprisingly, her new daughter-in-law has since bought an even more expensive Italian bedroom set.

The use of home space involves choice and norms of opulence; that one is at all in the position to make such choices implies a great deal about status as well. This, in turn, is reflected in the endless discussions about furniture and their cost, as well as the careful display of products such as fancy Italian bedroom sets. It is also reflected in the gaze of visitors as they enter a house and assess it in the terms of contemporary social norms.

The wealthier one is, the more one's domestic space is subject to visitors' gaze and assessment: bedrooms and bathrooms or even basements (usually set up as an extra apartment) are arranged as carefully as a museum exhibition. In the two dozen or so fancy homes in the town, visitors are routinely taken on complete tours that can include the master bedroom Jacuzzi, the teenage son's hi-tech bedroom in shades of gray, and the inside of the cherry wood linen closet. They are expected to ask questions about cost and style of furnishings, and to learn here the latest trends.

These spacious villas are those of entrepreneurs who have been successful in business, including at least two doctors and two lawyers with successful private practices. They contrast with the functional blocks of four apartments such as the one Sayideh and Hasan will build. Their inhabitants control interaction with extended family in a way they wouldn't if they were part of an extended household *hara* or building. While people are attracted to mansions as gathering places for important formal or informal gatherings, they are also kept at a distance by them, as the mansions reinforce class distinction and reify "Westernization." Within two generations, new class distinctions have arisen in Deir al-Asad, sometimes creating a distance between brothers and presumably their children. The image of the "nuclear" family, mirroring aspirations of "individualism" imputed to the West, can only be maintained if mansion owners can provide in the same way for their children—if they can manage to get land, build houses, and maintain or extend the social networks by which their relative value can be projected into the future.

One woman who moved into her villa in 1998 (they had begun building in 1988) commented that her children could now do their homework without being "disturbed" by the constant proximity of cousins who wanted to play. The mansion allowed her to control the comings and goings of the children by eliminating the obligation of constant informal sociability present in the extended family *hara* in which she had been living previously.

Another woman has recreated the social network she missed in her mansion's splendid isolation: she opened a beauty salon by which in 1998–1999 she introduced new punk hairstyles in the town. Her authority as a mansion owner allows her the latitude to set trends without being too vulnerable to gossip.

Sayideh and Hasan's piece of land, which lies against the sloping hillside, has a small hollow cave in it, about the size of a room. They have been thinking about what they might want to do with it, and have come up with an idea they like. They are going to turn it into an old-style Arab sitting room. They will put a curtain over the opening, and low pillows and carpets inside. There will be a charcoal brazier for the winter, and Narguilehs, which Hasan likes to smoke since Sayideh brought him one from Jordan, and good music. Maybe they will run an electrical cord there for light and music...

Interest in old things has become quite widespread in the Galilee in the 1990s. Old-fashioned objects are, of course, considered a part of one's identity and history; in that sense they are often also expressions of the nation inside the home. Into the late 1980s, national aspirations were illustrated through portraits of Nasser on living room walls, often flanked by the portraits of the man who had been the last *mukhtar* under the British, and had become a refugee in Lebanon in 1948. This man was known, among other things, for his nationalist views. In addition many walls sported maps of Palestine published in Lebanon and sold in Arab bookstores in the Galilee and the West Bank.

In the 1980s, the people who were interested in old objects were few, and passionate about it. In the late 1990s however, the past had become fashionable in a way it never had been before.[33] One Akka interior designer was selling "old-fashioned" interiors that had never existed, but his ideas were greatly in demand. He made imitation stone for columns and walls, imitation wood beams for ceilings, and imitation stone tiles for floors that could in turn be covered with expensive carpets. His customers were people who desired to have something with "character" in their homes, something "authentic."[34] Moreover, they were people who could afford it, established businessmen and professionals, people who saw themselves as self-confidently "Western" enough to encompass, claim, and display an "Arab character" in their living room. This display, of course, came with an offhand commentary listing the costs, lest anyone thought it was cheap. This was trend-setting, not, God forbid, regression. There was enough opulence and latest-model appliances in the house to make this point very clear. These inhabitants were not making a direct political claim over history, though Palestinian nationalism was considered to be an attribute and aspiration of all local political leaders; they were consuming "authenticity," and through it producing distinction. Yet it is striking that this style is making others think, more and more, about similar ideas.

Now it is acceptable for Sayideh and Hasan to conceive of their cave, and for other people to think of themselves as possibly coming to sit in it and enjoy it. The experience of previous generations in caves is not a consideration for them. The kind of refuge that Sayideh and Hasan are imagining is not that taken by shepherds, saints, refugees, or children; rather it is one that fits squarely within the modernity of their new home. What Sayideh and Hasan are thinking of putting together is not so much a place, but an experience, one that is enjoyable and worthwhile precisely because it can be shared with visitors and told in their accounts. After all, no one else in the town has a cave.

Sayideh and Hasan's new home, according to its size, layout, and projected furnishings, will stand for the achievement of a middle-income home in contemporary Deir al-Asad. It will represent their standing within local society, their own personal tastes, and their innovations. Yet even innovations, such as the cave, are shaped not only by the historical workings of desire, but also by the movement of power relations: what makes the idea of the cave possible to imagine for them is also the fact that it is acceptable in terms of the contemporary fashion of pastness and authenticity among well-to-do townspeople.

In order to try and understand the relation between consuming desires and social practices, the social meaning of furniture prices is especially important. Like most of the furniture in new homes I have visited, Sayideh and Hasan's will be overwhelmingly new and relatively expensive. The prices, like those of wedding furniture, will be made public, simultaneously signaling and defining the owners' current status. The fact that the family helped build the home, however, is so taken for granted that it will not be a topic of discussion. Likewise, Sayideh's earnings will be tactfully ignored, so as not to question what, according to current local norms, constitutes Hasan's authority as a head of household. Such contrasts between what is said, what is taken for granted, and what is silenced reveal a prominent, existing tension. This is a tension between defining one's household's position, meeting society's expectations for one's gender, class, and status, and acknowledging changing social relations. Together, these forces constitute the sociality of building; they also shape personal desires and aspirations.

Conclusion

Through a close look at the processes by which people fashion homes at the turn of the millennium, this chapter has provided an ethnographic exploration of the relation between state institutions, social practices, and personal desires. It also points to the importance of creative effervescence that brings people to re-imagine and refashion themselves, and possibly their relationship to others and institutions, through the not-so-simple project of making a home of their own.

The latest news provide one more instance of historical choice within political and economic constraints. In 2002, Sayideh and Hasan decided against the cave; they chose instead to carve extra space into the hillside to widen their home. This required additional loans (both from relatives and from Israeli financial institutions), additional work, and cuts in the family budget. They justified the greater expense by explaining that a larger home is a better long-term solution for the family's future—a future conceived, by the power of political constraints and historical circumstances that also shape imaginative desires, on the same piece of land.

Author's Note

The research on which this chapter is based was funded by grants from the MacArthur Council for the Advancement of Peace and International Cooperation (1989–90) and from the Lady Davies Foundation (1990–91). I thank Henya Rachmiel for her comments on an earlier draft.

Notes

1. Michel Foucault [Maurice Florence], "Foucault," *Dictionnaire des Philosophes*, 942–944 (Paris 1984). Quoted from http://foucault.info/foucault/biography.html.
2. Deir al-Asad is categorized as a "village" according to administrative definitions; local people describe it as *al-balad* (the village/homeland). I purposely use the word "town" for two reasons. First, the size of Deir al-Asad makes it comparable to small towns in other areas of the world. Second, and most importantly, I wish to avoid various connotations associated with the term "Arab village" in Israel. In my understanding, Deir al-Asad is not rural, since it has been the subject of in-situ urbanization like other villages around the world; in addition, most of its agri-cultural lands have been expropriated and as I note here, much of the lack of mobility of local people is due to state policies. Furthermore, the Deir al-Asad with which I am concerned here exists within power relations extant in contem-porary Israel rather than in a national imagination of a "backwards" past see Gil Eyal, "Between East and West: Discourse on 'the Arab Village' in Israel," *Teoria ve-Bikoret* 3 (1992): 39–54 (Hebrew).

3. See Arjun Appadurai, "Introduction," in Arjun Appadurai, ed., *The Social Life of Things* (Cambridge: Cambridge University press, 1986), 3–63; Jonathan Friedman, *Consumption and Identity* (Chur, Switzerland: Harwood Academic Publishers, 1994); Daniel Miller, *Acknowledging Consumption: A Review of New Studies* (London: Routledge, 1995).

4. Leora Auslander, *Taste and Power: Furnishing Modern France* (Berkeley: University of California Press, 1996); Victoria de Grazia, "Introduction" in Victoria de Grazia and Ellen Furlough, eds., *The Sex of Things: Gender and Consumption in Historical Perspective* (Berkeley: University of California Press, 1996).

5. Pierre Bourdieu, "The Kabyle House or the World Reversed," in *Algeria 1960: Essays*, trans. Richard Nice (Cambridge: Cambridge University Press, 1979), 133–153; Pierre Lévi-Strauss, *La Voie des Masques* (Paris: Plon, 1979).

6. Janet Carsten and Stephen Hugh-Jones, *About the House* (Cambridge: Cambridge University Press, 1995); Daniel Miller, *Home Possession: Material Culture Behind Closed Doors* (Oxford: Berg, 2001).

7. Ghazi Falah, "Israeli 'Judaization' Policy in Galilee and Its Impact on Local Arab Urbanization," *Political Geography Quarterly* 8, 3 (1989): 229–253; Ian Lustick, *Arabs in the Jewish State: Israel's Control of a National Minority* (Austin: University of Texas Press, 1980). Oren Yiftachel, *Planning a Mixed Region in Israel: The Political Geography of Arab–Jewish Relations in the Galilee* (London: Avebury–Gower, 1992).

8. Elsewhere I have argued that the notion of *dar* as a respectable extended household is also produced through making and using historical documents, in which some local sociopolitical power struggles and notions of history are inscribed. Tania Forte, "On Making a Village" (Ph.D. dissertation, University of Chicago, 2000).

9. Scott Atran "Demembrement Social et Remembrement Agraire dans un Village Palestinien," *L'Homme* 25, 4(96) (1985): 111–135.

10. Of these approximately 350 left for Lebanon, mostly settling in Ein al-Helweh refugee camp. Some of them and their descendants have come to visit over the years, and a few have married relatives in Deir al-Asad.

11. My evidence is partial, since there is no reference to the use of the concept in the nineteenth century. It does not appear in the archival documents from the Mandate period that I have perused.

12. Talal Asad, "Anthropological Texts and Ideological Problems: An Analysis of Cohen on Arab Villages in Israel," *Economy and Society* 4, 3 (1975): 247–282. See also Lustick, *Arabs in the Jewish State.*

13. Sabri Jiryis, *The Arabs in Israel* (Beirut: Institute of Palestine Studies, 1969); Hassan Amun et al., *Palestinians in Israel: Two Case Studies* (London: Ithaca Press, 1977); Avraham Haleli, "Rights in Land" in Avraham Shmueli, Aharon Sofer, and Norit Kliot, eds., *Artzot ha-Galil* (Haifa: Haifa University Press, 1984), 575–611 (Hebrew); Yiftachel, *Planning a Mixed Region.*

14. Granovsky [Granott] Abraham, *The Land Issue in Palestine* (Jerusalem: Keren Kayemet le-Israel, 1936).

15. Majid Al-Haj and Henri Rosenfeld, *Arab Local Government in Israel* (Haifa: University of Haifa Press, 1988).

16. In 1998, the exchange rate was fourteen to one according to the town engineer. See also Oren Yiftachel, *Guarding the Grove* (Beit Berl: The Institute for Israeli Arab Studies, 1997) (Hebrew).

17. See James Scott, *Seeing Like a State: How Certain Schemes to Improve on the Human Condition Have Failed* (New Haven: Yale University Press, 1998).

18. Sha`ab is an unusual town because it was emptied of its population in 1948, and repopulated with Arabs from other areas in the early 1950s. All the land there is considered state land, and it is available for long-term leases at relatively lower prices.

19. Note that although the situation is especially extreme in Palestinian areas, in the 1990s state officials maintained the price of land relatively high by controlling the amount of land released for building. By contrast, state officials offered land for settlement very cheap or even free in areas in which it aimed to attract a Jewish presence.

20. See also Rhoda Ann Kanaaneh, *Birthing the Nation: Strategies of Palestinian Women in Israel* (Berkeley: University of California Press, 2002).

21. Viviana Zelizer, *The Social Meanings of Money* (Princeton: Princeton University Press, 1997). Marx's famous discussion of the concept of value underlies much of theoretical thinking about the social and political implications of money.

22. Hilma Grandqvist, *Marriage Conditions in a Palestinian Village* (New York: AMS Press & Co., 1931).

23. See also David Graeber, "Beads and Money: Notes Toward a Theory of Wealth and Power," *American Ethnologist* 23, 1 (1996): 4–24.

24. In contrast to the assumption underlying Bourdieu's notion of social capital, I show here that Sayideh's actions are not meant to enhance her own self-interest—though they might indeed enhance her status. Instead, these actions had the potential to transform over time the value of the house and that of the jewelry. See also Hans Medick and David Sabean, eds., *Interest and Emotion: Essays on the Study of Family and Kinship* (Cambridge: Cambridge University Press, 1984).

25. See also Nancy Munn, *The Fame of Gawa: A Symbolic Study of Value Transformation in a Massim (Papua New Guinea) Society* (Cambridge: Cambridge University Press, 1986).

26. For an excellent politico-economic description of partnership contracts in the late nineteenth and early twentieth century, see Yaakov Firestone, "Crop-Sharing Economics in Mandatory Palestine, Part I," *Middle Eastern Studies*, 11, 1 (1975): 1–23; idem, "Crop-Sharing Economics in Mandatory Palestine, Part II," *Middle Eastern Studies* 11, 2 (1975): 175–194. For a description of the cultural and social meanings and practices involved, see Forte, "On Making a Village."

27. A major expense at that time is the dowry a boy's parents have to give to the girl's. Others are wedding expenses and the costs of setting up a new household.

28. Suad Amiry and Vera Tamari, *The Palestinian Village Home* (London: British Museum, 1989).

29. I came across this man at the local council office, and was presented to him as "a researcher." As a result, his account of local history differed somewhat from others I heard—he cautiously avoided any critical remarks about Israeli institutions.

30. See also Susan Slyomovics, *The Object of Memory: Jew and Arab Narrate the Palestinian Village* (Philadelphia: University of Pennsylvania Press, 1999).

31. Tsevet Tesh`a le-Tichnun Kollel, *Julis, Deir al-Asad, Nahf: Economic and Social Survey* (Tel Aviv: Segal Ba`am Press, 1969) (Hebrew).

32. Ibid., 16. At the same time, the state-initiated building projects to provide housing for thousands of Jewish immigrants, who were first housed in makeshift conditions. A typical apartment in a Jewish town had two or three rooms, or three to four rooms in rural areas. No housing projects were built for Arab communities.

33. While interest in history and memory may be the result of global trends see Andreas Huyssen, "Present Pasts: Media, Politics, Amnesia," *Public Culture* 12, 1 (Winter 1999: 21–38), its local manifestations are linked to specific patterns: an intense interest in the oral history of 1948, of one's family and village (see also Slyomovics, *The Object of Memory*; Forte, "On Making a Village"), and the development of Israeli tourism in the Galilee after the Oslo accords (Slyomovics, *The Object of Memory*); and the availability of specific means of inscription (especially websites, archives of Palestinian institutions and NGOs, and the wide availability of videocameras).

34. See also Johannes Fabian, *Time and the Other: How Anthropology Makes Its Objects* (New York: Columbia University Press, 1983).

CHAPTER 7

Consumption-Based Inequality: Household Expenditures and Possession of Goods in Israel, 1986–1998

Tally Katz-Gerro

Introduction

This chapter discusses consumption inequality in Israel in the period 1986–1998. I start out with a general discussion of the link between consumption and inequality and the different manifestations of this link. Next I describe the features of consumer culture in Israel. Finally, I analyze consumption inequality based on household expenditure categories and ownership of durable goods. I decompose changes in consumption patterns across three nationally representative surveys conducted in Israel, and I explore the way consumption patterns are associated with major social cleavages such as income, ethnicity, and social class.

Students of consumer society and consumer culture argue that elements of the system of production, such as occupation and class, are becoming less salient in defining the individual and the social structure, while consumption patterns are becoming the major determinants of individual identity and of the relationships between social groups. In consumer society we tend to think and speak of ourselves not as workers or citizens but as consumers. This is because consumption is not only a means to fulfill everyday needs, it is a central mechanism that organizes social relationships, social communication, and social hierarchy.

Two Approaches to Consumption and Social Inequality

Consumer culture promises a democratization of choice. Seemingly, discrimination on the basis of gender, race, or class does not exist at sites of consumption for as long as you have money you are a desired customer.[1] Even those who have limited financial resources have a choice among a variety of goods that suit their level of income and type of credit. Since consumption occurs in the market, laws of the market apply to create a feeling of freedom and ultimate individuality that is attained by commodities. But in practice, the contention that consumer society promotes a democracy of consumers is very problematic.[2] Although in theory choice increases or maximizes, in practice there is evidence that consumption patterns reflect dimensions of inequality and therefore help retain power relations and social inequality in two ways. First, economic resources like income, access to credit (and the different steps in a credit hierarchy: platinum, gold, silver), and occupation stratify consumption options. Second, consumer culture promotes hierarchies of style, taste, and culture that are constantly created and changed. These hierarchies both reconstitute and challenge social cleavages that are based on ethnicity, gender, class, age, and education.

To understand the link between material consumption and inequality we should ask whether inequality in consumption reflects existing dimensions of inequality or creates new ones. This question, which pertains to the nature of the relationship between the social matrix and consumption patterns, receives two different answers. According to one approach, certain taste profiles and consumption preferences are associated with social classes and produce distinctions in a certain time and place.[3] That is, social groups reproduce the boundaries between them by using different consumption styles. According to the second approach, since consumer culture allows for a multiplicity of identities, cultural meanings, and carriers of communication, we cannot find a simple correlation between consumption styles and existing social categories. Consumption styles do not reproduce boundaries between social categories but draw new boundaries between *cultural tribes*.[4] These tribes include, for example, Internet users, smokers, non-smokers, green consumers, palm pilot users, anti-consumers, soccer team fans, bridge players, and more. These cultural tribes are not detached from a context of class, education, or ethnicity, but their common denominator and the symbolic capital they produce are consumptive.

It is not the purpose of this chapter to adjudicate the debate between the two approaches but to acknowledge that both have merit. Incorporating these approaches into one theoretical framework allows us to examine how consumption both accentuates configurations of social stratification and creates new dynamics of inequality. The sphere of consumption outlines dimensions that create new hierarchies by which people are measured and compete and by which resources are distributed and inequality is constituted. Consumer culture—as a social arrangement in which the relation between lived culture and social resources, between meaningful ways of life and the symbolic and material resources on which they depend, is mediated through markets[5]—frames, contextualizes, and justifies these processes. The act of consumption has a deep social cultural meaning that goes beyond economic terms. Therefore, the complexity of the link between consumption and inequality needs to be examined through the intertwined links between economy and culture, with an emphasis on the reproduction of existing dimensions of inequality on the one hand, and the creation of new dimensions on the other hand.

Consumption and Reproduction of Existing Dimensions of Inequality

Let me now elaborate on the approach that maintains that inequality in consumption patterns reflects first and foremost inequality in other factors such as income, education, and ethnicity. I choose to follow in this direction because (as will become evident later) the type of data I use lends itself to this type of analysis. At the same time, the general theoretical framework encompasses both directions discussed in the previous section.

Extensive research discusses the nature and direction of the association between consumption patterns, income, and education.[6] Consumption patterns play an important role in defining and challenging social hierarchies like race,[7] or gender.[8] Since consumption styles and tastes are socially stratified and structured and are not merely the result of individual whim, they constitute a resource that social groups use in establishing and perpetuating their position in the social structure. Processes of socialization in the family and in school determine taste in a wide range of areas such as food, clothing, home decoration, hobbies, and art. Everyday consumption habits are important in maintaining the basic structures of power and inequality.[9]

Along with the association between inequality and individual consumption an association exists between inequality and consumption at the level of the state. The idea of collective consumption emphasizes the role of the state and its relationship to capital, to the allocation of resources, and to the promotion of different interests in the relation between consumption and material, cultural, and social reproduction.[10] An analysis of the interaction of private and public provision, welfare policy, and regulation of consumption levels presents a complex link between stratification and consumption patterns.[11] Realms that traditionally existed outside the market—health care, education, the arts, elderly care—are brought into the market and turn into an additional means of increasing sales and profit. Therefore, consumption patterns reproduce elements of social stratification, both through socio-demographic characteristics and through the structural location of indivi-duals or groups within the state. The stratification of consumers to "consumption classes" affects lifestyles, attitudes, and status, and creates in turn a differential advantage in access to consumption. The centrality of consumption has implications for social inequality, both from the viewpoint that stresses the differential status of goods and the way they create social distinction and from the viewpoint that maintains a correlation between consumer culture and manipulation of cultural codes and symbols.

Consumption and Inequality in Israel

Israel has exhibited many of the characteristics of a consumer society especially in the past ten years. We see this in the development of shopping centers on the outskirts and shopping malls in the cities; in the increase in the number of retail chains (American and others); in the increase in private consumption; in the development of e-commerce, on-line shopping, and a shopping channel on cable TV; in the strengthening link between religious holidays and shopping carnivals; in the multiplicity of consumerism columns in the media; and in the activities of consumer organizations. Consumer cul-ture in Israel is also evident in the development of a politics of consumption. We witness public struggles over the outcomes of consumption, its effect on environmental issues, and the effect of consumer culture on the societal fabric.[12] Consumer culture in Israel is also present in the realm of cultural consumption. Research furnishes reports that social groups such as classes and ethnic groups are stratified according to their tastes and cultural con-sumption preferences,[13] and according to their material possessions,[14] and that leisure and recreational activities are becoming more and more signifi-cant to Israelis in terms both of time and the importance attributed to them.[15] These features make an investigation of the link between consump-tion and inequality in Israel a relevant and timely issue.

Research Questions

Household consumption expenditure and the acquisition of durable goods is a topic of interest among researchers in a variety of disciplines. A review of the literature indicates that the perspectives of economists (durable good forecasts), sociologists (household decision behavior), and psychologists (individual consumer behavior) have contributed to the body of knowledge regarding household consumption behavior.[16] Past studies in Israel have mostly been concerned with isolated, fragmentized aspects of economic behavior. This study takes a more general approach in addressing household expenditure and its allocation to various categories such as food, housing, education, and health. The chapter attempts to shed light on the answers to two main questions: first, can we find systematic variance in consumption patterns of households in terms of the relative shares of different expenditure categories and possession of durable goods? Second, which socioeconomic variables are significant predictors of variation in consumption patterns? Additional goals of the chapter are to identify changes over time in household consumption patterns in the period 1986–1998; and to explore the relative importance that households under budget constraints give to different commodity categories by assuming that the allocation of expenditures indicates different consumption priorities.

Data

Data are drawn from three household expenditure surveys conducted by the Israeli Central Bureau of Statistics in 1986/1987, 1992/1993, and 1997/1998. This choice of time framework stems from considerations of data availability.[17] The surveys are nationally representative and use samples drawn from all households in urban localities in Israel. The data sets describe the characteristics of households' standard of living by obtaining components of their budget, consumption patterns, level and composition of income, and housing conditions. The unit of analysis in these surveys is the household: a group of people living in the same dwelling most days of the week with a shared expenditure budget for food. The 1986/1987 survey covers 5,000 households, the 1992/1993 survey covers 5,212 households, and the 1997/1998 survey covers 5,862 households.

Results

Table 7.1 describes the variables that stand at the center of the analysis, namely expenditure categories. Each main household expenditure

Table 7.1 Detailed components of consumption expenditure categories

Food	Housing	Dwelling	Furniture and household equipment	Clothing and footwear	Health	Education, culture, entertainment	Transport and communication
Bread	Government taxes	Water, gas, electricity	Furniture	Outerwear	Health insurance	Education services	Public transport
Cereal, pastry	Monthly rent	Maintenance and renovation	Electrical equipment	Underwear	Dental treatment	Newspapers, books, stationery	Travel abroad
Vegetable oils and products	Housing consumption in kind	Domestic help	Non-electrical equipment	Footwear	Expen. on health services	Culture, sport, entertainment	Expen. on vehicles
Meat and poultry	Other (insurance etc.)	Household articles	Bedding and towels	Sewing, fabrics, accessories	Other expen. on health	Recreation and excursions	Other expen.
Fish		Municipal taxes	Home decoration	Cleaning and laundry outside home		Entertainment durable goods	Post, telephone, communication
Milk, milk products, eggs						Hobbies, sport, camping equipment	
Sugar, sugar products							
Soft drinks							
Alcoholic beverages							
Meals away from home							
Misc. food products							

category—food; housing; dwelling; furniture and household equipment; clothing and footwear; health; education; culture and entertainment; transport and communication—consists of detailed categories that are specified in the table.

The average percentage of each expenditure category as part of the total household budget is depicted in table 7.2, which also compares percentage expenditure on various categories in thirteen other countries. Expenditures on food, on clothing and footwear, on housing equipment, and on transport and communication in Israel are seen to be comparable to most other countries; expenditure on housing in Israel is relatively high and similar to that in Denmark and Sweden; expenditure on health is relatively low due to a national health insurance scheme; expenditure on education and culture is relatively higher than in other countries; and finally, expenditure on miscellaneous is a heterogeneous category and therefore difficult to interpret.[18]

After reviewing the main expenditure categories and their distribution, I now turn to analyzing the longitudinal dimension. Figure 7.1 provides an

Table 7.2 A cross-national comparison of the structure of household budgets (in % of total household expenditures)*

	Food	Clothing and footwear	Housing	Housing equipment	Health and medical care	Transport and communication	Education and culture	Misc.
Germany	15.1	7.1	19.6	8.5	15.1	15.3	9.2	10.1
Denmark	20.8	5.2	28.8	6.1	2.2	15.4	10.4	11.1
Spain	10.0	8.1	13.0	6.5	4.7	15.3	6.6	25.8
France	18.0	5.9	20.6	7.3	10.0	15.5	7.3	15.5
Greece	36.4	7.7	13.5	7.4	4.2	14.7	5.3	10.9
Italy	20.2	9.1	16.9	9.1	7.1	11.6	8.8	17.2
Netherlands	14.8	6.8	19.0	6.9	13.1	12.6	10.2	16.6
UK	20.6	5.9	19.5	6.6	1.7	17.1	10.2	18.3
Sweden	19.9	5.8	32.9	6.6	2.3	15.7	9.5	7.2
Norway	21.6	6.7	24.4	6.2	2.6	16.6	9.6	12.3
Canada	15.7	5.2	24.7	8.7	4.6	14.3	11.2	15.6
USA	11.4	5.9	18.1	5.8	17.8	14.0	10.3	16.5
Japan	19.9	5.8	20.8	5.9	11.3	9.7	10.7	16.0
Israel	18.0	3.7	31.7	5.3	4.0	18.8	13.8	4.7

* Data for Israel is from 1997/1998; data for all other countries is from 1997.
Source: Luis Cases, *La Consommation des Ménages en 1997* (Paris: INSEE, 1999)
Central Bureau of Statistics publication no. 1147. *Household Expenditure Survey 1997–1998: General Summary.*

overview of change over time in consumption practices in Israel. We see that household monthly consumption expenditure as a percentage of total expenditure has decreased in the categories of food, furniture and equipment, clothing, and health. The percentage of expenditure on housing, transport and communication, and education and culture has increased. Expenditure on household maintenance has remained stable. These changes reflect changing consumption habits, changing income profiles, and changes in price indices.

The most important expenditure group in 1986/1987 was food, having a budget share of 17 percent. There was a relative decrease in 1997/1998 when the food share was 14 percent. On the other hand, the shares of housing (23 percent) and transport and communication (19 percent) increased significantly from 1986/1987. Such patterns as found here are common in societies in fairly rapid economic growth.

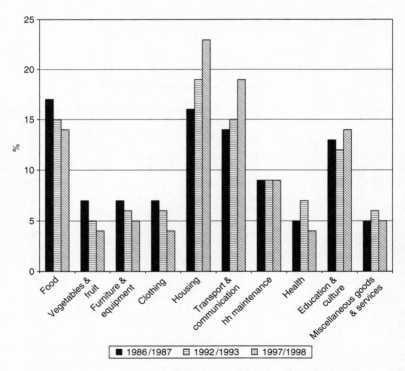

Figure 7.1   Household monthly consumption expenditure by year.

These results become more significant when linked to the concept of "priority," reflecting the importance that households attribute to consumption areas. The concept of priority has been employed for understanding the process by which households acquire durable goods.[19] Evidence from these studies demonstrates the existence of some underlying priority pattern or order in which household durable goods are purchased. The concept of priority can be linked to theories of practice that emphasize the routine nature of conduct and the importance of practical decision making in everyday life. Practices are routinized types of behavior, understanding, knowhow, and desire; they represent a pattern of unique actions.[20] Consumption occurs within practices and analysis of routinized consumption reveals the patterning of practices.

Priority patterns of household expenditures are demonstrated in figures 7.2, 7.3, and 7.4, which depict a rank order of consumption expenditure categories by income deciles, each figure representing a different survey year.

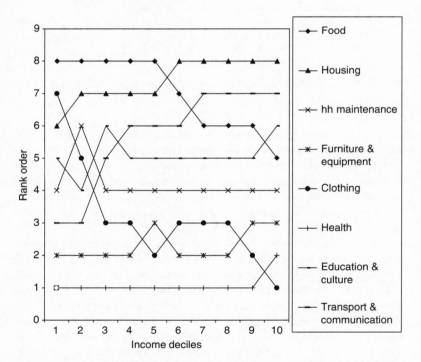

Figure 7.2   Rank order of consumption expenditure by income deciles, 1986/1987.

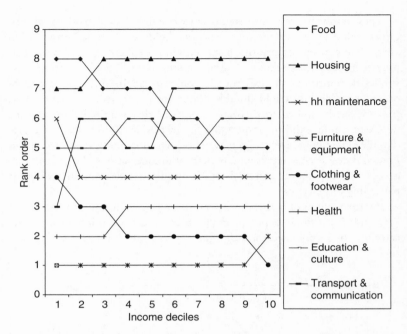

Figure 7.3   Rank order of consumption expenditure by income deciles, 1992/1993.

I used the relative expenditure on each category (food, housing, dwelling, etc.) to order the categories from lowest expenditure to highest.[21] I then ranked the relative expenditures for each income decile separately. Each figure shows the rank order of each category within a specific income decile; in other words the priority of expenditures and differences in priority order between deciles. A comparison of the three figures provides a picture of change over time.

For example, in figure 7.2, the Y-axis represents the order of eight consumption categories according to the relative size of expenditure. The X-axis represents ten income deciles from the lowest (one) to the highest (ten). Individuals in the lowest income decile in 1986/1987 spent on average the least on health, more than that on furniture and equipment, more than that on transport and communication, and the like. They spent the highest relative expenditure on food. The priority order for the second decile looks different. Households belonging to this decile spent the least on health and the most on food, just like households of the first decile, but they spent

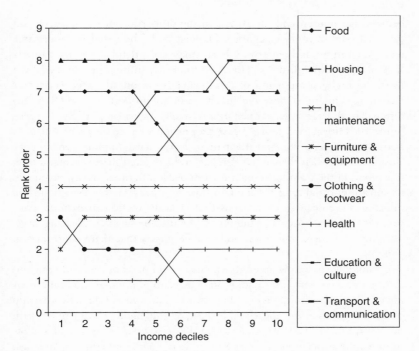

Figure 7.4   Rank order of consumption expenditure by income deciles, 1997/1998.

a larger share of their expenditure on household maintenance and on housing, and they spent a smaller share on clothing and education and culture.

Expenditure categories of food and housing receive a high rank in all three surveys. In 1986/1987 and 1992/1993, food is more important than housing for the lower deciles. Transport and communication also receive a high rank, which rises still higher for the upper deciles. Over time, transport and communication become more important, and in 1997/1998 the eighth, ninth, and tenth deciles rank this category at the top in terms of expenditure share. Next comes education and culture. Household maintenance is stable in the middle, after which come clothing and footwear, and furniture and equipment. Health starts from lowest rank in 1986/1987, then moves up to the fourth or fifth position in 1992 and goes down to first or third position in 1997/1998.

Over all, the ordering of the categories is more or less stable across income deciles. This is surprising because we would expect to find differences between the poorest and the richest in the proportion of income spent on different consumption categories. Douglas and Isherwood propose three types of consumption patterns.[22] First, small scale, defined by high proportion of total expenditure on food. Second, medium scale, defined by a relatively higher proportion of total expenditure on advanced consumer technology combined with a relatively declining proportion of total expenditure. Third, large scale, defined by a relatively higher proportion of expenditure on information and a lesser proportion of income spent on food. Surprisingly, we do not find this pattern in the data. Another general conclusion is that when there is a difference in the ordered share of expenditure, it is found in the eighth, ninth, and tenth deciles. In other words, inequality in consumption among different income groups is mainly evident in the upper income deciles. This means that class distinction in material consumption in Israel occurs at the top of the income hierarchy, leaving large, seemingly homogeneous lower and middle classes that share consumption patterns.

I would like to follow up on the discussion of household expenditures and income brackets with information on ownership of durable goods and their association with ethnic origin.[23] Figures 7.5, 7.6, and 7.7 depict ownership of durable goods in the household by head of household's continent of birth, in 1986/1987, 1992/1993, and 1997/1998 respectively. These figures allow us to see the association between ownership of goods and ethnicity in Israel. Most Jewish households have a refrigerator, oven, washing machine, color TV, and a telephone. The other appliances are less prevalent. In 1992/1993, there was an increase in the proportion of households owning an air conditioner, VCR, and personal computer. The bar that represents non-Jews is always lower than the bars representing other ethnic groups.

Differences between Jews and non-Jews significantly decreased between 1986 and 1998, probably due to massive immigration of non-Jews from the former Soviet Union during the 1990s but also due to the expansion of consumer society. This is particularly evident in ownership of deep freezer, washing machine, color TV, telephone, and car. On most items, households headed by Israeli-born individuals have an advantage. In other words, we see here that advantages that originate in classical determinants of social inequality, such as ethnicity, are reinforced and reflected in advantages in the realm of consumption.

Interestingly, several durable goods mark a distinction or demarcate a difference between Jews (Israeli, Ashkenazi, Mizrachi) and non-Jews (mostly Arabs), while other goods mark a distinction among all ethnic groups, as represented in figures 7.8–7.11. Refrigerator, oven, washing machine, TV,

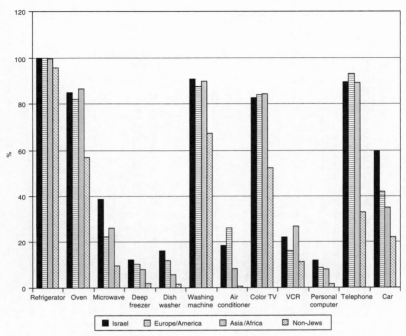

Figure 7.5 Ownership of durable goods by continent of birth of head of household, 1986/1987.

telephone, and car differentiated Jews and non-Jews in the surveys of 1986/1987 and 1992/1993 but no longer in 1997/1998. Microwave, dishwasher, air conditioner, VCR, and personal computer still differentiate the two groups in 1997/1998. A difference seems still to exist between Jewish and non-Jewish households mainly in terms of convenience electrical appliances, and information and communication appliances. We find a similar picture when it comes to differences among ethnic groups in terms of possessions. Microwave, dishwasher, VCR, personal computer, and car are the main possessions that marked differences between ethnic groups in the period 1986–1998.

The results presented thus far lead the way to a more statistically rigorous analysis of the relationship between factors of inequality and consumption patterns, in the form of a multivariate regression model. Such a model considers the simultaneous effects of various independent variables on the explained variable, in this case expenditure categories. A regression model uses a linear equation to quantify the relationship between the socio-demographic

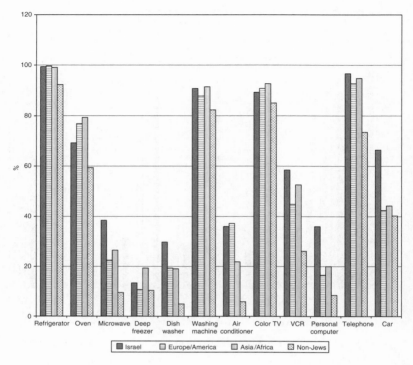

Figure 7.6 Ownership of durable goods by continent of birth of head of household, 1992/1993.

variables and the expenditure variable. The relationship between the variables is expressed in two different, but related, aspects: whether a significant association exists between each independent variable (e.g., income) and the dependent variable (expenditure category) and the strength of this association. To accommodate data peculiarities and to simplify the presentation, I applied the multivariate analysis only to the most recent survey of 1997/1998.

Table 7.3 provides results of analysis of the variables affecting proportionate consumption expenditure (each expenditure category is calculated as a proportion of the total consumption expenditure for each household). The numbers presented in the table represent the variables that have a statistically significant impact on the various expenditure categories. Empty cells in table 7.3 represent insignificant effects on the expenditure categories.

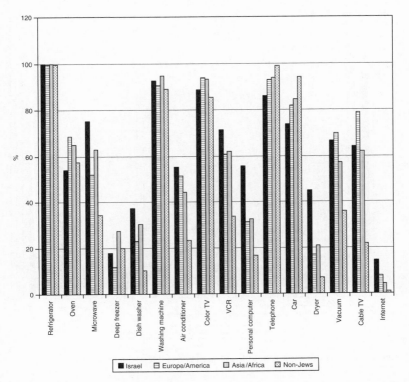

Figure 7.7 Ownership of durable goods by continent of birth of head of household, 1997/1998.

A comparison of the significance of the socioeconomic variables across expenditure categories reveals the determinants of expenditures.

The socioeconomic variables included in the model are more successful in explaining variance in the expenditures on food, housing, and education and culture. Family size, age, and years of schooling of the head of household seem to be the main variables that significantly affect most expenditure categories. Housing and education and culture seem to be the main expenditure categories that are affected by the head of household's class position. All classes spend proportionately more than the upper class on housing, and proportionately less than the upper class on education and culture. This evinces a clear link between inequality and stratification in the sphere of production and inequality in the sphere of consumption. At the same time, class position seems to have an occasional effect on the other categories, meaning that class distinction occurs mainly in housing and education and culture

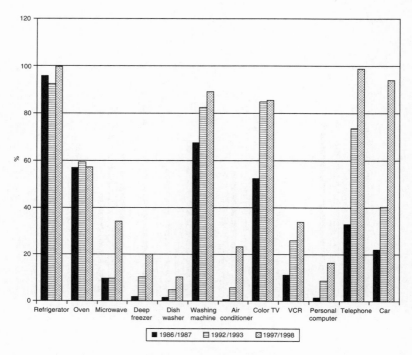

Figure 7.8 Ownership of durable goods, 1986–1998, non-Jews only.

(which are major life outcomes) and is absent from other dimensions of consumption, at least in terms of the measures employed here.

Immigration status of heads of household matters only regarding new immigrants (immigrated between 1990 and 1998). This is interesting because it is an indication of the level of integration of this group, and it means that those who immigrated to Israel before 1990 have adopted consumption patterns that are similar to the majority of Israelis.

As is pointed out in figures 7.8–7.11, differences among ethnic groups are not statistically significant in the 1997/1998 survey. According to table 7.3, Ashkenazi respondents spend more than Israeli-born (the reference category) on housing, there are no differences between Mizrachi and Israeli-born on any of the expenditure categories, and non-Jews spend more than Israeli-born on food and clothing and less on health and on education and culture.

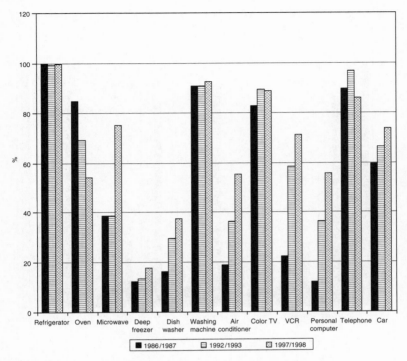

Figure 7.9 Ownership of durable goods, 1986–1998, Israeli-born only.

Discussion

The results discussed here provide an overview of inequality in consumption in Israel over a period of one and a half decades. Consumption patterns are measured on the basis of household expenditures and possession of durable goods. One of the contributions of this study is its exploration of the way in which consumption patterns are linked to other aspects of social stratification. The basic question addressed by the multivariate analysis is whether systematic differences exist in households' consumption patterns and how these differences are associated with traditional dimensions of inequality, specifically income, family size, gender, age, education, immigration status, ethnicity/nationality, and class. The results suggest that consumption-based stratification both echoes and crosscuts other social cleavages.

As suggested here, the first main finding is that income does not always explain variation in consumption. It is therefore important to ask what

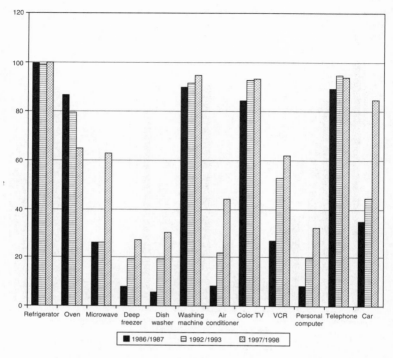

Figure 7.10 Ownership of durable goods, 1986–1998, Mizrachi only.

accounts for disparities in consumption that are not explained by income. We observe overall differential expenditure by family size, age, years of schooling of head of household, immigration status, and ethnicity (Jews vs. non-Jews). Occupational class position of head of household seems to have significant influence on expenditure in some of the categories but not in others. An interesting conclusion here is that ethnicity seems to matter less than class position, contrary to previous analyses of Israeli society that tend to emphasize the ethnic cleavage over the class cleavage. In 1997/1998 we see that consumption patterns are not demarcated by ethnicity but they are demarcated by ethnicity to a certain extent.

The results exhibit the structural link between social position and material resources that are typical of a stratified society. Advantages that originate in classical determinants of social inequality are associated with advantages in the realm of consumption. At the same time, not all expenditure categories

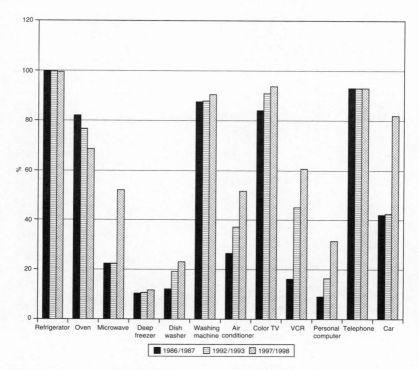

Figure 7.11 Ownership of durable goods, 1986–1998, Ashkenazi only.

and not all nondurable goods mark a distinction between ethnic groups or between classes. These findings exemplify the complex way that the economy is embedded within the social—household consumption is embedded within socioeconomic factors. Moreover, it shows how economic (priority profiles) and social (ethnicity, class, age) categories are merged within the structures of market relations and microeconomic action.

A comparison of the priority pattern of households within different income brackets reveals that the ordering of expenditure categories is quite stable across income deciles. This indicates that income is not the most significant factor that divides social groups in terms of consumption patterns. Most of the difference between income deciles is located at the top of the income hierarchy, namely in the three upper deciles. In terms of change over time, it seems that differences between income brackets and between ethnic groups diminished from the first to the last survey. Consumption inequality still exists but it is more attenuated over time.

Table 7.3 Significant coefficients of ordinary least squares regressions on proportional expenditure categories, 1997/1998 (reference categories: nonimmigrant, Israeli-born, and professionals and managers)

	Food	Housing	Dwelling	Furniture	Clothing	Health	Education, culture
Income	−.106	−.166			−.034	−.033	
Family size	.188	−.204	.035	−.060		−.080	.149
Male	.049	−.063	−.031				
Age	.033	.113	.142	−.055	−.207	.188	−.121
Years schooling	−.151	−.049	−.071	−.043		.039	.088
Immigrant new[a]	.182	.031	−.135	.081			−.112
Immigrant old[b]							−.047
Ashkenazi Mizrachi		.051					
Non-Jew	.186				.234	−.078	−.275
Class: Sales, service		.035	−.043				−.047
Skilled, farm		.046					
Unskilled		.050					−.032
Not employed		.100			.070	.064	−.062
Adjusted R²	.17	.17	.05	.01	.10	.07	.13
N	5790	5774	5746	4342	4760	4951	5722

[a] Immigrated between 1990 and 1998.
[b] Immigrated before 1990.

The issues discussed in this chapter open up numerous research questions to be pursued in the future. First, these results represent a series of measurement and methodology decisions made by the statistics bureau that collected the data. Therefore, the analysis and conclusions depend on the indicators, questions, and items chosen by the data collectors, and are not necessarily ideal in terms of addressing questions on the link between consumption and inequality. Future research may find it worthwhile to examine motivations and changes in the categories used by statistical offices when collecting and classifying information on consumption patterns.

Second, the role of the state and state agencies in regulating, changing, and shaping private and collective consumption is an important aspect of the relationship between consumption and inequality. Change over time in state intervention is particularly important for understanding variation in consumption patterns. For example, changes in national health laws alter the minimum and maximum expenditure of households on this category and the association between expenditure on health and factors such as income. In another example, changes in the housing market and its regulation affect changes in the distribution of household expenditure on rent, mortgage, loans, and the like.

Finally, a more detailed analysis of the components of each expenditure category could provide additional fine-tuned conclusions regarding areas of consumption that are associated with differences between socio-demographic variables. Such a detailed analysis could also focus on differences between material consumption (housing, dwelling) and cultural consumption (education, culture) thus encompassing a wide range of consumer behavior.

Despite the rising interest in consumer society among social scientists, there has been scant examination of this topic in Israel. Hopefully, the issues discussed in this chapter will open up an exchange on the unique features of consumer society in Israel both in terms of the heavy influence of European and American culture and in terms of its geopolitical location in the Middle East.

Author's Note

I would like to thank Relli Shechter for his vision and initiative in organizing an international workshop on "Considering Consumption, Production, and the Market in the Constitution of Meaning in the Middle East and Beyond" at Ben-Gurion University in Spring 2001. I thank the participants of the workshop for a lively discussion and insightful comments. This chapter also benefited from comments made by participants in the workshop on economic sociology that took place at the University of Haifa in 2002. Finally, I thank Sivan Shohat for her research assistance and Relli Shechter for his comments on the final version of the chapter. Please address correspondence to Tally Katz-Gerro, Department of Sociology and Anthropology, University of Haifa, Mount Carmel, Haifa 31905, Israel. E-mail: tkatz@soc.haifa.ac.il

Notes

1. Mary Douglas and Baron Isherwood, *The World of Goods: Towards an Anthropology of Consumption* (London: Routledge, 1979); Robert Sack, *Place, Modernity, and the Consumer's World* (Baltimore: Johns Hopkins Press, 1992).
2. Don Slater, *Consumer Culture and Modernity* (Cambridge: Polity, 1997); Celia Lury, *Consumer Culture* (New Brunswick: Rutgers University Press, 1996).
3. Thorstein Veblen, *The Theory of the Leisure Class* (New York: Viking Press, 1931 [1899]); Paul DiMaggio, "Cultural Entrepreneurship in Nineteenth-Century Boston, I: The Creation of an Organizational Base for High Culture in America,"

Media, Culture, and Society 4 (1982): 33–50; Pierre Bourdieu, *Distinction: A Social Critique of the Judgement of Taste*, trans. Richard Nice (London: Routledge and Kegan Paul, 1984).

4. Lury, *Consumer Culture*; Michel Mafessoli, *The Time of the Tribes: The Decline of Individualism in Mass Society* (London: Sage, 1996); Jan Pakulski and Malcolm Waters, *The Death of Class* (London: Sage, 1996); Slater, *Consumer Culture and Modernity*.

5. Slater, *Consumer Culture and Modernity*, 8.

6. Douglas and Isherwood, *The World of Goods*; Richard Peterson and Roger Kern, "Changing Highbrow Taste: From Snob to Omnivore," *American Sociological Review* 61, 5 (1996): 900–907; Tally Katz-Gerro, "Highbrow Cultural Consumption and Class Distinction in Italy, Israel, West Germany, Sweden, and the United States," *Social Forces* 81, 8 (2002): 207–229; Erik Bihagen, "How Do Classes Make Use of Their Incomes?" *Social Indicators Research* 47 (1999): 119–151; Martin Lux, "Changes in Consumption of Households during 1990–1997," *Czech Sociological Review* 8, 2 (2000): 211–232.

7. Virág Molnár and Michèle Lamont, "How Blacks Use Consumption to Shape Their Collective Identity: Evidence from African American Marketing Specialists," *Journal of Consumer Culture* 1, 1 (2001): 31–46.

8. Erik Bihagen and Tally Katz-Gerro, "Culture Consumption in Sweden: The Stability of Gender Differences," *Poetics* 27, 5, 6 (2000): 327–349.

9. Bourdieu, *Distinction*.

10. Manuel Castells, *The Urban Question: A Marxist Approach* (London: Edward Arnold, 1977).

11. Ben Fine, "From Political Economy to Consumption," in *Acknowledging Consumption: A Review of New Studies*, ed. Daniel Miller (London: Routledge, 1995), 127–163.

12. Some notable examples are: the campaign against a new cross-country toll motorway; consumer boycott on products produced in the territories occupied by Israel in 1967; boycott by orthodox Jews on products that do not bear certain seals of kosher food; campaign for recycling of glass and plastic bottles; and campaigns against the opening of various McDonald's branches.

13. Tally Katz-Gerro and Yossi Shavit, "The Stratification of Leisure and Taste: Classes and Lifestyles in Israel," *European Sociological Review* 14, 4 (1998): 369–386.

14. Moshe Semyonov, Noah Lewin-Epstein, and Seymor Spilerman, "The Material Possessions of Israeli Ethnic Groups," *European Sociological Review* 12, 3 (1996): 289–301.

15. Bihagen and Katz-Gerro, "Culture Consumption in Sweden."

16. Semyonov et al., "The Material Possessions"; Susse Georg, "The Social Shaping of Household Consumption," *Ecological Economics* 28 (1999): 455–466. Lux, "Changes in Consumption of Households."

17. Since 1999, the Israeli Central Bureau of Statistics has conducted yearly surveys that pertain to household expenditure.

18. Additional cross-national comparative information is available on the publication of books, consumption of cultural goods and services, production of films, cultural activities, cultural practices and heritage, cultural trade and communication, and cultural trends at http://www.unesco.org/culture/worldreport/html_eng/tables2.shtml.

19. R.F. Lusch, E.F. Stafford, and J.J. Kasulis, "Durable Accumulation: An Examination of Priority Patterns," in *Advances in Consumer Research* 5 (1978): 119–125, ed. H. Keith Hunt, J. Kasulis, R. Lusch, and E. Stafford, "Consumer Acquisition Patterns for Durable Goods," *Journal of Consumer Research* 6 (1979): 47–57; P.R. Dickson, R. Lusch, and W. Wilkie, "Consumer Acquisition Priorities for Home Appliances: A Replication and Re-evaluation," *Journal of Consumer Research* 9, 4 (1983): 432–435.

20. Pierre Bourdieu, *The Logic of Practice*, trans. Richard Nice (Cambridge: Polity, 1992); Andreas Reckwitz, "Toward a Theory of Social Practices: A Development in Culturalist Theorizing," *European Journal of Social Theory* 5, 2 (2002): 243–263.

21. By analyzing consumption expenditure shares instead of absolute expenditures, income loses its dominant position so that the relevance of other explanatory variables can be examined.

22. Douglas and Isherwood, *The World of Goods*.

23. Israeli Jews consist of about three main ethnic groups (and several subgroups according to country of origin): Israeli-born Jews, Mizrachi Jews who originate from Asia and Africa, and Ashkenazi Jews who originate from Europe and America.

Bibliography

`Abd al-Karim, Ahmad `Izzat. *Tarikh al-ta`lim fi `asr muhammad `ali.* Cairo: Maktabat al-Nahda al-Misriyya, 1939.

Abu-Lughod, Janet. *Before European Hegemony: The World System A.D. 1250–1350.* New York: Oxford University Press, 1989.

———. *Cairo: 1001 Years of the City Victorious.* Princeton: Princeton University Press, 1971.

Abu-Lughod, Lila, ed. *Remaking Women: Feminism and Modernity in the Middle East.* Princeton: Princeton University Press, 1998.

Aciman, André. *Out of Egypt.* New York: Riverhead Books, 1994.

Addison, James Thayer. *The Christian Approach to the Moslem.* New York: Columbia University Press, 1942.

Aglietta, Michel. *A Theory of Capitalist Regulation: The US Experience.* Translated by David Fernbach. London: New Left Books, 1979.

Ahmed, Leila. *Women and Gender in Islam.* New Haven: Yale University Press, 1992.

Allen, Roger. *A Study of Hadith `Isa Ibn Hisham: Muhammad al-Muwaylihi's View of Egyptian Society during the British Occupation.* Albany: State University of New York Press, 1974.

Allen, Roger. *The Arabic Novel: An Historical and Critical Introduction.* Syracuse: Syracuse University Press, 1982.

Allen, William. "Sixty-five Istanbul Photographers, 1887–1914." In *Shadow and Substance: Essays in the History of Photography,* edited by Kathleen Collins, 127–136. Bloomfield Hills: Amorphous Institute Press, 1990.

Almog, Oz. *The Sabra: A Profile.* Tel Aviv: Am Oved, 2001. (Hebrew).

Amin, Qasim. *Al-Mar'a al-jadida.* Cairo, 1900.

———. *Tahrir al-mar'a.* Cairo, 1899.

Amiry, Suad and Vera Tamari. *The Palestinian Village Home.* London: British Museum, 1989.

Amun, Hassan et al. *Palestinians in Israel: Two Case Studies.* London: Ithaca Press, 1987.

Anderson, Benedict. *Imagined Communities.* London: Verso, 1991.

Andrews, Maggie J. and Mary M. Talbot, eds. *All the World and Her Husband: Women in Twentieth-Century Consumer Culture.* London: Cassell, 2000.

Anonymous. *Arkitekt* 14, 11–12 (1944): 278–283.

——. "Ensiegnement de la géographie en Egypte," *Buletin de la societé de géographie a Paris*; Deuxiéme série, 3 (1835).

——. *Kadın* April 25 (1959): n.p.

Appadurai, Arjun, ed. *The Social Life of Things: Commodities in Cultural Perspective.* Cambridge: Cambridge University Press, 1986.

Aronson, Amy Beth. *Taking Liberties: Early American Women's Magazines and Their Readers.* New York: Praeger, 2002.

Artin, Yacub. "Essai sur les causes de renchérissement de la vie materielle au Caire au courant du XIXe siecle." In *The Economic History of the Middle East, 1800–1914: A Book of Readings* edited by Charles Issawi, 449–451. Chicago: University of Chicago Press, 1966.

——. *L'Instruction Publique en Egypte.* Paris, 1890.

Asad, Talal. "Anthropological Texts and Ideological Problems: An Analysis of Cohen on Arab Villages in Israel," *Economy and Society* 4, 3 (1975): 247–282.

Asdar, Ali Karman. *Planning the Family in Egypt: New Bodies, New Selves.* Austin: University of Texas Press, 2002.

Atay, Falih Rifki. *Çankaya: Atatürk devri hatıraları.* Istanbul: Dünya Yayınları, 1958.

Atran, Scott. "Demembrement Social et Remembrement Agraire dans un Village Palestinien," *L'Homme* 25, 4:96 (1985): 111–135.

Auslander, Leora. *Taste and Power: Furnishing Modern France.* Berkeley: University of California Press, 1996.

`Awad, Louis. *Tarikh al-fikr al-misri al-hadith, min al-hamla al-Faransiyya ila `asr isma`il.* Cairo: Maktabat Madbouli, 1987.

Babazogli, Sophie. "l'Education de la jeune fille musalmane en Egypt." Thesis, l'Ecole des Hautes Etudes Sociales, Paris, 1927, Cairo: Paul Barbey, 1928.

Ballaster, Ros et al. *Women's Worlds: Ideology, Feminity and the Woman's Magazine.* London: Macmillan, 1991.

Baron, Beth. "The Making and Breaking of Marital Bonds in Modern Egypt." In *Women in Middle Eastern History: Shifting Boundaries in Sex and Gender*, edited by Nikki Keddie and Beth Baron, 275–291. New Haven: Yale University Press, 1991.

——. *The Women's Awakening in Egypt.* New Haven: Yale University Press, 1994.

Bar-Yosef, Rivkah. "Household Management in Two Types of Families in Israel: Applying an Organizational Model to Comparative Analysis." In *Families in Israel*, edited by Shamgar-Handelman, Leah and Rivkah Bar-Yosef, 169–196. Jerusalem: Academon, 1991. (Hebrew).

Baudrillard, Jean. *The System of Objects.* Translated by James Benedict. London: Virgo Press, 1996.

Bauer, Arnold J. *Goods, Power, History: Latin America's Material Culture.* Cambridge: Cambridge University Press, 2001.

Beaugé, Gilbert and Engin Çizgen. *Images d'Empire. Aux origines de la photographie en Turquie/Türkiye'de fotoğrafın öncüleri.* Istanbul: Institut d'études françaises d'Istanbul (n.d.).

Beetham, Margaret. *A Magazine of Her Own? Domesticity and Desire in the Woman's Magazine, 1800–1914*. London: Routledge, 1996.

Behrens-Abouseif, Doris. *Azbakiyya & Its Environs from Azbak to Ismail, 1876–1879*. Cairo: L'Institut Français D'Archeologie Orientale, 1985.

Berkey, Jonathan. "Women and Islamic Education in the Mamluk Period." In *Women in Middle Eastern History: Shifting Boundaries in Sex and Gender*, edited by Nikki Keddie and Beth Baron, 143–157. New Haven: Yale University Press, 1991.

Bianchi, N.A. "Catalogue général des livres Arabes, Persans et Turcs imprimés à Boulac en Egypte depuis l'introduction de l'imprimerie dans ce pays," *Nouveau Journal Asiatique* 2 (1843): 24–60.

Bihagen, Erik. "How Do Classes Make Use of Their Incomes?" *Social Indicators Research* 47 (1999): 119–151.

—— and Tally Katz-Gerro. "Culture Consumption in Sweden: The Stability of Gender Differences," *Poetics* 27 5/6 (2000): 327–349.

Birdwell-Pheasant, Donna and Denise Lawrence-Zuniga. *House Life: Space, Place and Family in Europe*. Oxford: Berg, 1999.

Birenbaum-Carmeli, Daphna. *Tel Aviv North: The Making of a New Israeli Middle Class*. Jerusalem: Magnes Press, 2000. (Hebrew).

Bourdieu, Pierre. *Distinction: A Social Critique of the Judgement of Taste*. Translated by Richard Nice. London: Routledge and Kegan Paul, 1984.

——. "The Kabyle House or the World Reversed." *In Algeria 1960: Essays*. Translated by Richard Nice, 133–153. Cambridge: Cambridge University Press, 1979.

——. *The Logic of Practice*. Translated by Richard Nice. Cambridge: Polity, 1992.

Boyer, Robert. "The Eighties: The Search for Alternatives to Fordism." In *The Politics of Flexibility: Restructuring State and Industry in Britain, Germany, and Scandinavia* edited by Bob Jessop et al., 106–132. Aldershot, UK: Edward Elgar, 1991.

Brewer, John and Roy Porter. *Consumption and The World of Goods*. London: Routledge, 1993.

Broderick, Mary et al., eds. *A Handbook for Travellers to Lower and Upper Egypt*. London, 1896.

Bryden, Inga and Janet Floyd, eds. *Domestic Space: Reading the Interior in Nineteenth Century Britain and America*. Manchester: Manchester University Press, 1999.

Buğra, Ayşe and G. Irzık. "Human Needs, Consumption and Social Policy," *Economics and Philosophy* 15 (1999): 187–208.

——. "The Immoral Economy of Housing in Turkey," *The International Journal of Urban and Regional Research* 22 (1998): 303–317.

——. "Non-Market Mechanisms of Market Formation: The Development of the Consumer Durables Industry in Turkey," *New Perspectives on Turkey* 19 (1998): 29–52.

Burke, Timothy. *Lifebuoy Men, Lux Women: Commodification, Consumption, and Cleanliness in Modern Zimbabwe*. Durham: Duke University Press, 1996.

Camaroff, John and Jean L. Camaroff, *Ethnography and the Historical Imagination*. Boulder: Westview Press, 1992.

Campbell, Colin. *The Romantic Ethic and the Spirit of Modern Consumerism.* Oxford: Blackwell, 1987.

Carrier, James G. *Gifts and Commodities: Exchange and Western Capitalism since 1700.* London: Routledge, 1995.

Carsten, Janet and Stephen Hugh-Jones. *About the House: Lévi-Strauss and Beyond.* Cambridge: Cambridge University Press, 1995.

Cases, Luis. *La Consommation des Ménages en 1997.* Paris: INSEE, 1999.

Caspi, Dan and Yehiel Limor. *The Mediators: The Mass Media in Israel, 1949–1990.* Tel Aviv: Am Oved, 1992. (Hebrew).

Castells, Manuel. *The Urban Question: A Marxist Approach.* London: Edward Arnold, 1977.

Cecil, Edward. *The Leisure of an Egyptian Official.* London: Hodder & Stoughton, 1921.

Central Bureau of Statistics. *Household Expenditure Survey.* Various issues. (Hebrew).

Chapman, Tony and Jenny Hockey. *Ideal Homes? Social Change and Domestic Life.* London: Routledge, 1999.

Cieraad, Irene. *At Home: An Anthropology of Domestic Space.* Syracuse: Syracuse University Press, 1999.

Çizgen, Engin. *Photographer Ali Sami, 1866–1936.* Istanbul: Haşet Kitabevi, 1989.

———. *Photography in the Ottoman Empire, 1839–1919.* Istanbul: Haşet Kitabevi, 1987.

Clarke, Alison J. *Tupperware: The Promise of Plastic in 1950's America.* Washington: Smithsonian Institution Press, 1999.

Cohen-Avigdor, Nava. *"Female Politicians (vis a vis Male Politicians) in the Israeli Women's Press: Representations During Election Years 1959, 1977, 1996,"* M.A. thesis, Bar Ilan University, 1998. (Hebrew).

Cohen, Olga. "Models of Womanhood as Extracted from Hebrew Popular Magazines of the 1930s and 1940s in the Yishuv Culture in Palestine," M.A. thesis, Tel-Aviv University. In preparation. (Hebrew).

Cole, Juan R.I. *Colonialism and Revolution in the Middle East: Social and Cultural Origins of Egypt's `Urabi Movement.* Princeton: Princeton University Press, 1993.

Cole, Sam and Ian Miles. *Worlds Apart: Technology and North–South Relations in the Global Economy.* Brighton: Wheatsheaf Books, 1984.

Crabbs, Jack A. *The Writing of History in Nineteenth-Century Egypt: A Study in National Transformation.* Cairo: The American University in Cairo Press, 1984.

Cuno, Ken. "Ambiguous Modernization: The Transition to Monogamy in the Khedival House of Egypt." Paper presented at the Annual Meeting of the Middle East Studies Association, San Francisco, November 18, 2001.

Davis, Deborah S. *The Consumer Revolution in Urban China.* Berkeley: University of California Press, 2000.

De Grazia, Victoria with Ellen Furlough. *The Sex of Things: Gender and Consumption in Historical Perspective.* Berkeley: University of California Press, 1996.

Delanoue, Gilbert. *Moralistes et Politiques Musulmans dans l'Egypte du XIXéme siécle*, Vol. 2. Cairo: Institut Français d'Archéologie Orientale du Caire, 1982.

De Leon, Edward. *The Khedive's Egypt*. London, 1882 [1877].

Depping, Georges-Bernard. *Aperçu historique sur les moeurs et coutumes des nations*.... Paris, 1826.

Deringil, Selim. *The Well-Protected Domains: Ideology and the Legitimation of Power in the Ottoman Empire, 1876–1909*. London: I.B. Tauris, 1998.

Dickson, P.R., R. Lusch and W. Wilkie. "Consumer Acquisition Priorities for Home Appliances: A Replication and Re-evaluation," *Journal of Consumer Research* 9, 4 (1983): 432–435.

DiMaggio, Paul. "Cultural Entrepreneurship in Nineteenth-Century Boston, I: The Creation of an Organizational Base for High Culture in America," *Media, Culture, and Society* 4 (1982): 33–50.

Douglas, Mary and Baron Isherwood. *The World of Goods: Towards an Anthropology of Consumption*. London: Routledge, 1979.

Doumani, Beshara, ed. *Family History in the Middle East: Household, Property, and Gender*. Albany: State University of New York Press, 2003.

Duben, Alan and Cem Behar. *Istanbul Households: Marriage, family and fertility 1880–1940*. Cambridge: Cambridge University Press, 1991.

Du Gay, Paul and Michael Pryke. *Cultural Economy: Cultural Analysis and Commercial Life*. London: Sage, 2002.

El-Azhari Sonbol, Amira, ed. *Women, the Family, and Divorce Laws in Islamic History*. Syracuse: Syracuse University Press, 1996.

Elgat, Zvi and David Paz. *'Isha [Woman] 2000: The Stories, the Dramas, the Style and the Beauty in Israel, 1948–2000*. Tel Aviv: Yedi'ot Akhronot, 1999. (Hebrew).

El-Gawhary, Mahmoud. *Ex-Royal Palaces in Egypt*. Cairo: Dar al-Maarif, 1954.

Ethnos 67, 3 (2002). A special issue on consumption in Eastern Europe.

Even-Zohar, Itamar."Factors and Dependencies in Culture: A Revised Draft for Polysystem Culture Research," *Canadian Review of Comparative Literature* 24, 1 (1997): 15–34.

——. "The Making of Culture Repertoire and The Role of Transfer," *Target* 9, 2 (1997): 373–381.

Eyal, Gil. "Between East and West: Discourse on 'the Arab Village' in Israel," *Teoria ve-Bikoret* 3 (1992): 39–54. (Hebrew).

Fabian, Johannes. *Time and the Other: How Anthropology Makes Its Objects*. New York: Columbia University Press, 1983.

Fahmy, Khaled. *All the Pasha's Men: Mehmet Ali, his Army, and the Making of Modern Egypt*. Cambridge: Cambridge University Press, 1997.

Falah, Ghazi. "Israeli 'Judaization' Policy in Galilee and its Impact on Local Arab Urbanization," *Political Geography Quarterly* 8, 3 (1989): 229–253.

Farid, Zaynab. *Ta`lim al-mar'a al-`arabiyya fi-l-turath wa-fi-l-mujtama`at al-`ara-biyya al-mu`asira*. Cairo: Maktabat al-Anglo-al-Misriyya, 1980.

Fay, Mary Ann. *"Women and Households: Gender, Power, and Culture in Eighteenth Century Egypt."* Ph.D. dissertation, Georgetown University, 1993.

Felix, David. "Interrelations between Consumption, Economic Growth, and Income Distribution in Latin America since 1800." In *Consumer Behavior and Growth in the Modern Economy*, edited by Henry Baudet and Henk van de Meulen, 133–177. London: Croom Helm, 1982.

Ferguson, Marjorie. *Forever Feminine: Women's Magazines and the Cult of Femininity*, Aldershot, UK: Gower, 1983.

Fine, Ben. "From Political Economy to Consumption." In *Acknowledging Consumption: A Review of New Studies*, edited by Daniel Miller, 127–163. London: Routledge, 1995.

—— and Ellen Leopold. *The World of Consumption*. London: Routledge, 1993.

Finney, Minnehaha. "Kindergarten Schools in Egypt, the Vision." In *The Child in the Midst: The Story of the Beginning and Development of the Kindergarten in Mission Schools and in the Government Schools of Egypt*. n.p./n.d.

Firestone, Yaakov. "Crop-Sharing Economics in Mandatory Palestine—Part I." *Middle Eastern Studies* 11, 1 (1975): 1–23.

——. "Crop-Sharing Economics in Mandatory Palestine—Part II." *Middle Eastern Studies* 11, 2 (1975): 175–194.

Forte, Tania. "On Making a Village". Ph.D. dissertation, University of Chicago, 2000.

Fox-Genovese, Elizabeth. *Within the Plantation Household*. Chapel Hill: University of North Carolina Press, 1988.

Foucault, Michel (Maurice Florence). "Foucault," *Dictionnaire des Philosophes*, 942–944 (Paris, 1984). Quoted from http://foucault.info/foucault/biography.html.

Friedman, Jonathan, ed. *Consumption and Identity*. Chur, Switzerland: Harwood Academic Publishers, 1994.

Al-Gamayyil, Antun. *Al-Fatat wa-l-bayt*. Cairo: Matba`at al-Ma`arif, 1916.

Garvey, Ellen Gruber. *The Adman in the Parlor: Magazines and the Gendering of Consumer Culture, 1880s to 1910s*. Oxford: Oxford University Press, 1996.

Gavin, Carney E.S., ed. "Imperial Self-Portrait: The Ottoman Empire as Revealed in the Sultan Abdul Hamid II's Photographic Albums," *Journal of Turkish Studies* 12, (1988).

Gelpi, Rosa-Maria et François Julien-Labruyère. *Histoire du crédit a la consommation: Doctrines et pratiques*. Paris: Découverte, 1994.

Georg, Susse. "The Social Shaping of Household Consumption," *Ecological Economics* 28 (1999): 455–466.

Gillette, Maris Boyd. *Modernization and Consumption among Urban Chinese Muslims*. Stanford: Stanford University Press, 2000.

Godlewska, Anne and Neil Smith. *Geography and Empire*. Oxford: Blackwell Publishers, 1994.

Göçek, Fatma Müge. *East Encounters West: France and the Ottoman Empire in the Eighteenth Century*. New York: Oxford University Press, 1987.

——. *Rise of the Bourgeoisie, Demise of Empire: Ottoman Westernization and Social Change*. New York: Oxford University Press, 1996.

Göksu, Sezai. "Yenişehir'de Bir İmar Öyküsü," İlhan Tekeli (der.), *Kent, Planlama, Politika ve Sanat: Tarık Okyay Anısına Yazılar,* 257–276. Ankara: ODTÜ Mimarlık Fakültesi, 1994.

Graeber, David. "Beads and Money: Notes Toward a Theory of Wealth and Power," *American Ethnologist* 23, 1 (1996): 4–24.

Gran, Peter. *Islamic Roots of Capitalism, 1769–1840.* Austin: University of Texas Press, 1979.

Grandqvist, Hilma. *Marriage Conditions in a Palestinian Village.* New York: AMS Press & Co., 1931.

Granovsky [Granott], Abraham. *The Land Issue in Palestine.* Jerusalem: Keren Kayemet le-Israel, 1936.

Grey, Mrs. William. *Journal of a Visit to Egypt, Constantinople, the Crimea, & c. in the Suite of the Prince and Princess of Wales.* New York, 1870.

Gripsrud, Jostein, ed. *Television and Common Knowledge.* London: Routledge, 1999.

Al-Haj, Majid and Henri Rosenfeld. *Arab Local Government in Israel.* Haifa: University of Haifa Press, 1988.

Haleli, Avraham. "Rights in Land." In *Artzot ha-Galil*, edited by Avraham Shmueli, Aharon Sofer and Norit Kliot, 575–611. Haifa: Haifa University Press, 1984. (Hebrew).

Harvey, David. *The Condition of Postmodernity: An Enquiry into the Origins of Cultural Change.* Oxford: Blackwell, 1989.

Hathaway, Jane. *The Politics of Households in Ottoman Egypt: The Rise of the Qazdağlis.* Cambridge: Cambridge University Press, 1997.

Helman, Anat. "The Development of Civil Society and Urban Culture in Tel Aviv during the 1920s and 1930s." Ph.D. dissertation, Hebrew University, Jerusalem, 2000. (Hebrew).

Herzog, Hanna. "The Women's Press in Israel: An Arena for Reproduction or Challenge?" *Kesher* 28 (2000): 36–42. (Hebrew).

Heyworth-Dunne, John. *An Introduction to the History of Education in Modern Egypt.* London: Frank Cass, 1968.

Howes, David, ed. *Cross-Cultural Consumption: Global Markets, Local Realities.* London: Routledge, 1996.

Hunter, Robert F. *Egypt Under the Khedives, 1805–1879: From Household Government to Modern Bureaucracy.* Pittsburgh: The University of Pittsburgh Press, 1984.

Huyssen, Andreas. "Present Pasts: Media, Politics, Amnesia," *Public Culture* 12, 1 (Winter, 1999): 21–38.

Ilbert, Robert. *Heliopolis, Le Caire, 1905–1922: Genése d'une Ville.* Paris: Centre Nationale de la Recherche Scientifique, 1981.

Islamoglu-Inan, Huri, ed. *The Ottoman Empire and the World-Economy.* Cambridge: Cambridge University Press, 1987.

Jackson, Peter, Michelle Lowe, Daniel Miller and Frank Mort. *Commercial Cultures: Economies, Practices, Spaces.* London: Berg, 2000.

Jiryis, Sabri. *The Arabs in Israel.* Beirut: Institute of Palestine Studies, 1969.

Joseph, Suad, ed. *Intimate Selving in Arab Families: Gender, Self, and Identity.* Syracuse: Syracuse University Press, 1999.

—— and Susan Slymovics, eds. *Women and Power in the Middle East.* Philadelphia: University of Pennsylvania Press, 2001.

Kabbani, Rana. *Europe's Myths of Orient.* London: Quartet Books, 1986.

Kanaaneh, Rhoda Ann. *Birthing the Nation: Strategies of Palestinian Women in Israel.* Berkeley: University of California press, 2002.

Karl, Rebecca E. "Creating Asia: China in the World at the Beginning of the Twentieth Century," *American Historical Review* 103, 4 (1998): 1096–1118.

Kasaba, Reşat. *The Ottoman Empire and the World Economy: The Nineteenth Century.* Albany: State University of New York Press, 1988.

Kasulis, J., R. Lusch and E. Stafford. "Consumer Acquisition Patterns for Durable Goods," *Journal of Consumer Research* 6 (1979): 47–57.

Katz-Gerro, Tally. "Highbrow Cultural Consumption and Class Distinction in Italy, Israel, West Germany, Sweden, and the United States," *Social Forces* 81, 8 (2002): 207–229.

—— and Yossi Shavit. "The Stratification of Leisure and Taste: Classes and Lifestyles in Israel," *European Sociological Review* 14, 4 (1998): 369–386.

Keddie, Nikki and Beth Baron, eds. *Women in Middle Eastern History: Shifting Boundaries in Sex and Gender.* New Haven: Yale University Press, 1991.

Keleş, Ruşen "Housing Policy in Turkey." In *Housing Policy in Developing Countries*, edited by Gil Shidlo, 140–172. London: Routledge, 1990.

Keren, Michael. "The Woman and Civil Society in Eretz Israel during the 1920s," *Kesher* 28 (2000): 36–42. (Hebrew).

Kern, Roger. "Changing Highbrow Taste: From Snob to Omnivore." *American Sociological Review* 61, 5 (1996): 900–907.

Klein, Naomi. *No Space, No Choice, No Jobs, No Logo: Taking Aim at the Brand Bullies.* New York: Picador USA, 1999.

Kopytoff, Igor. "The Cultural Biography of Things: Commoditization as Proces." In *The Social Life of Things: Commodities in cultural perspective*, edited by Arjun Appaduri, 64–91. Cambridge: Cambridge University Press, 1986.

Laden, Sonja."Domesticity and Representations of Self and Reality, in *la-'Isha*, an Israeli Women's Weekly," *Kesher* 28 (2000): 36–42. (Hebrew).

——. "Magazine Matters: Toward a Cultural Economy of the South African (Print) Media." In *Media, Democracy and Renewal in Southern Africa*, Keyan Tomaselli and Hopeton Dunn, eds., 181–208. International Academic Publishers: Denver, 2000.

——. " 'Making the Paper Speak Well', or, the Pace of Change in Consumer Magazines for Black South Africans," *Poetics Today* 22, 2 (2001): 515–548.

——. "Middle-Class Matters, or, How to Keep Whites Whiter, Colors Brighter, and Blacks Beautiful," *Critical Arts* 11, 1/2 (1997): 120–41.

Lane, Edward. *An Account of the Manners and Customs of the Modern Egyptians.* 2 Vols. London, 1846.

Lash, Scott and John Urry. *Economies of Signs and Space*. London: Sage, 1994.

Lee, Martyn J. *Consumer Society Reborn: The Cultural Politics of Consumption*. London: Routledge, 1993.

Lerner, Daniel. *The Passing of Traditional Society: Modernizing the Middle East*. New York: Free Press of Glencoe, 1964.

Lévi-Strauss, Pierre. *La Voie des Masques*. Paris: Plon, 1979.

Lewis, Bernard. *The Emergence of Modern* Turkey. London: Oxford University Press, 1965.

Liebes, Tamar and Elihu Katz. *The Export of Meaning: Cross-Cultural Readings of Dallas*. New York: Oxford University Press, 1990.

Lipietz, Alain. *Mirages and Miracles: The Crises of Global Fordism*. Translated by David Macey. London: Verso, 1987.

Livingstone, Sonia and Peter Lunt. *Talk on Television: Audience Participation and Public Debate*. London: Routledge, 1994.

Loeb, Lori Anne. *Consuming Angels: Advertising and Victorian Women*. New York: Oxford University Press, 1994.

Lourca, Anwar. *L'Or de Paris*. Paris: Sindbad Press, 1988.

——. *Voyageurs et écrivains Egyptiens en France au vingtiéme siécle*. Paris: Didier Press, 1970.

Lury, Celia. *Consumer Culture*. New Brunswick: Rutgers University Press, 1996.

Lusch, R.F., E.F. Stafford and J.J. Kasulis. "Durable Accumulation: An Examination of Priority Patterns," *Advances in Consumer Research* 5 (1978): 119–125. Edited by H. Keith Hunt.

Lustick, Ian. *Arabs in the Jewish State: Israel's Control of a National Minority*. Austin: University of Texas Press, 1980.

Lux, Martin. "Changes in Consumption of Households during 1990–1997," *Czech Sociological Review* 8, 2 (2000): 211–232.

Mabro, Judy. *Veiled Half-Truths: Western Travellers' Perceptions of Middle Eastern Women*. London: I.B. Tauris, 1991.

Mafessoli, Michel. *The Time of the Tribes: The Decline of Individualism in Mass Society*. London: Sage, 1996.

Majdi, Salah. *Hilyat al-zaman bi-manaqib khadim al-watan. Sirat rifa`a rafa`i al-tahtawi*. Edited by Jamal al-Din al-Shayyal. Cairo: Wizarat al-Thaqafa wa-l-Irshad al-Qawmi, 1958.

Makdisi, Jean Said. "Teta, Mother, and I." In *Intimate Selving in Arab Families: Gender, Self, and Identity*, edited by Suad Joseph, 25–52. Syracuse: Syracuse University Press, 1999.

Makhlouf Obermeyer, Carla. *Family, Gender, and Population in the Middle East: Policies in Context*. Cairo: American University in Cairo Press, 1995.

Malte-Brun, Conrad. *System of Universal Geography, Containing a Description of all the Empires, Kingdoms, States and Provinces in the Known*. Translated by James Percival. Boston, 1834.

Marchand, Roland. *Advertising the American Dream: Making Way for Modernity*. Berkeley: University of California Press, 1985.

Marcus, Sharon. *Apartment Stories: City and Home in Nineteenth-Century Paris and London.* Berkeley: University of California Press, 1999.

Marling, Karal Ann. *As Seen on TV: Visual Culture of Everyday Life in the 1950s.* Cambridge, MA: Harvard University Press, 1994.

Marsot, Afaf Lutfi al-Sayyid. *Egypt in the Reign of Muhammad `Ali.* Cambridge: Cambridge University Press, 1984.

———. "Revolutionary Gentlewomen in Egypt." In *Women in the Muslim World,* edited by Lois Beck and Nikki Keddie, 261–276. Cambridge, MA: Harvard University Press, 1978.

McCracken, Grant D. *Culture and Consumption: New Approaches to the Symbolic Character of Consumer Goods and Activities.* Bloomington: Indiana University Press, 1988.

McKendrick, Neil, John Brewer and J.H. Plumb. *The Birth of a Consumer Society: The Commercialization of Eighteenth-Century England.* Bloomington: Indiana University Press, 1982.

Medick, Hans and David Sabean, eds. *Interest and Emotion: Essays on the Study of Family and Kinship.* Cambridge: Cambridge University Press, 1984.

Meriwether, Margaret L. *The Kin Who Count: Family and Society in Ottoman Aleppo, 1770–1840.* Austin: University of Texas Press, 1999.

Micklewright, Nancy. *A Victorian Traveler in the Middle East: The Travel Writing and Photograph Albums of Annie Lady Brassey.* London: Ashgate, 2003.

———. "London, Paris, Istanbul, and Cairo: Fashion and International Trade in the 19th Century," *New Perspectives on Turkey* 7 (1992): 125–136.

———. "Musicians and Dancing Girls: Images of Women in Ottoman Painting." In *Women in the Ottoman Empire: Middle Eastern Women in the Early Modern Era,* edited by Madeline C. Zilfi, 153–168. Leiden: E.J. Brill, 1997.

———. "Negotiating between the Real and the Imagined: Portraiture in the Late Ottoman Empire." In *M. Uğur Derman Armağanı/M. Uğur Derman Feshschrift,* edited by Irvin Cemil Schick, 417–438. Istanbul: Sabancı Universitesi, 2000.

———. "Personal, Public and Political (Re)Constructions: Photographs and Consumption." In *Consumption Studies and the History of the Ottoman Empire, 1500–1922,* edited by Donald Quataert, 261–288. Albany: State University of New York Press, 2000.

———. "*Women's Dress in Nineteenth Century Istanbul: Mirror of a Changing Society,*" Ph.D. dissertation, University of Pennsylvania, 1986.

Mikha'il, Francis. *Al-Nizam al-manzili.* Cairo: Matba`at al-Ma`arif, 1913.

———. *Al-Tadbir al-manzili al-hadith.* 2 parts. Cairo: Matba`at al-Ma`arif, 1910.

Miller, Daniel, ed. *Acknowledging Consumption: A Review of New Studies.* London: Routledge, 1995.

———, ed. *Home Possessions: Material Culture Behind Closed Doors.* Oxford: Berg, 2001.

———. *Material Culture and Mass Consumption.* Oxford: Blackwell, 1987.

Mitchell, Timothy. *Colonising Egypt.* Cambridge: Cambridge University Press, 1988.

Molnár, Virág and Michèle Lamont. "How Blacks Use Consumption to Shape their Collective Identity: Evidence from African American Marketing Specialists," *Journal of Consumer Culture* 1, 1 (2001): 31–46.

Mortaş, Abidin. "Ankara'da Mesken Meselesi," *Arkitekt* 13, 11/12 (1943): 239–240.

Mubarak, `Ali. *Al-Khitat al-tawfiqiyya al-jadida li-misr al-qahira wa-mudunha wa- biladha al-qadima wa-l-shahira.* 2 Vols. Cairo: Dar al-Kutub, 1969.

Munn, Nancy. *The Fame of Gawa: A Symbolic Study of Value Transformation in a Massim (Papua New Guinea) Society.* Cambridge: Cambridge University Press, 1986.

Al-Muwaylihi, Muhammad. *Hadith `isa ibn hisham aw fatra min zaman.* 4th ed. Cairo: Dar al-Qawmiyya al-Tafaa wa-l-Nashr, 1964.

Najmabadi, Afsaneh. "Crafting an Educated Housewife in Iran." In *Remaking Women: Feminism and Modernity in the Middle East,* edited by Lila Abu-Lughod, 91–125. Princeton: Princeton University Press, 1998.

Naor, Mordechai. *Book of the Century.* Tel Aviv: Am Oved, 1996. (Hebrew).

Necipoğlu, Gülru et al. *The Sultan's Portrait: Picturing the House of Osman.* Istanbul: Işbank, 2000.

Nelson, Nina. *Shepheard's Hotel.* London: Barrie and Rockliff, 1960.

Newcomb, Horace and Paul M. Hirsch. "Television as a 'Cultural Forum.'" In *Television: The Critical View,* edited by Horace Newcomb, 503–515. New York: Oxford University Press, 1994.

Nord, David Paul. "A Republican Literature: Magazine Reading and Readers in Late-Eighteenth-Century New York." In *Reading in America,* edited by Cathy N. Davidson, 114–139. Baltimore: The Johns Hopkins University Press, 1989.

Offen, Karen. "Liberty, Equality, and Justice for Women: The Theory and Practice of Feminism in Nineteenth Century Europe." In *Becoming Visible: Women in European History,* edited by Renate Bridental, Claudia Koontz and Susan Stuard 335–373. 2nd ed. Boston: Houghton Mifflin, 1987.

Offer, Avner. "Why has the Public Sector Grown so Large in Market Societies? The Political Economy of Prudence in the UK, c.1870–2000," *Discussion Papers in Economic and Social History.* University of Oxford, 44, March 2002.

Öncü, Ayşe. "The Politics of Urban Land Market in Turkey: 1950–1980," *International Journal of Urban and Regional Research* 12 (1988): 38–64.

Orlove, Benjamin, ed. *The Allure of the Foreign: Imported Goods in Postcolonial Latin America.* Ann Arbor: University of Michigan Press, 1997.

Owen, Roger. *The Middle East in the World Economy, 1800–1914.* London: I.B. Tauris, 1993.

Pakulski, Jan and Malcolm Waters. *The Death of Class.* London: Sage, 1996.

Pamuk, Şevket. *The Ottoman Empire and European Capitalism, 1820–1913: Trade, Investment, and Production.* Cambridge: Cambridge University Press, 1987.

Peirce, Leslie P. *The Imperial Harem: Women and Sovereignty in the Ottoman Empire.* New York: Oxford University Press, 1993.

Perez, Nissan. *Focus East: Early Photography in the Near East (1839–1885).* New York: Abrams, 1988.

Peterson, Richard and Roger Kern. "Changing Highbrow Taste: From Snob to Omnivore," *American Sociological Review* 61, 5 (1996): 900–907.

Polanyi, Karl. "Economy as Instituted Process." In *Trade and Market in the Early Empire: Economies in History and Theory*, edited by K. Polanyi, Conrad M. Arensberg and Harry W. Pearson, 243–270. Chicago: Gateway, 1957.

———. *The Great Transformation.* Boston: Beacon Press, 1944.

———. *The Livelihood of Man.* New York: Academic Press, 1977.

Pollard, Lisa. "The Family Politics of Colonizing and Liberating Egypt, 1882–1919," *Social Politics* 7, 1 (2000): 47–79.

Posner, Ze'ev., "The Flat and the Street in Israel as a Component of World Modelling," M.A. thesis, Tel Aviv University, 1998. (Hebrew).

Pulat, G. *Dar Gelirlilerin Konut Sorunu ve Soruna Mekansal Çözüm Arayışları.* Ankara: Kent Koop, 1992.

Quataert, Donald, ed. *Consumption Studies and the History of the Ottoman Empire, 1550–1922: An Introduction.* Albany: State University of New York Press, 2000.

Radner, Hilary. *Shopping Around: Feminine Culture and the Pursuit of Pleasure.* London: Routledge, 1995.

Rappaport, Erika. *Shopping for Pleasure: Women in the Making of London's West End.* Princeton: Princeton University Press, 2000.

Razin, Assaf and Efraim Sadka. *The Economy of Modern Israel: Malaise and Promise.* Chicago: Chicago University Press, 1993.

Raymond, André. *Cairo.* Translated by Willard Wood. Cambridge, MA: Harvard University Press, 2000.

———. "Le Caire Sous Les Ottoman (1517–1798)." In *Palais et Maisons du Caire, Vol. II, Epoque Ottomane (XVI–XVIIIéme siécles)*, edited by Bernard Maury, André Raymond, Jacques Revault and Mona Zakariyya. Paris: Editions du Centre National de la Recherche Scientifique, 1983.

Reckwitz, Andreas. "Toward a Theory of Social Practices: A Development in Culturalist Theorizing," *European Journal of Social Theory* 5, 2 (2002): 243–263.

Reid, Susan E. and David Crowley, eds. *Style and Socialism: Modernity and Material Culture in Post-War Eastern Europe.* London: Berg, 2000.

Ritzer, George. *The McDonaldization of Society: An Investigation into the Changing Character of Contemporary Social Life.* Newbury Park, CL: Pine Forge Press, 1993.

Russell, Mona. "Competing, Overlapping, and Contradictory Agendas: Egyptian Education under British Occupation, 1882–1922." *Comparative Studies of South Asia, Africa, and the Middle East*, forthcoming.

———. "Creating *al-Sayyida al-Istihlakiyya*: Advertising in Turn-of-the-Century Egypt." *Arab Studies Journal* 8, 2/9, 1 (Fall 2000, Spring 2001): 61–96.

Sack, Robert. *Place, Modernity, and the Consumer's World.* Baltimore: Johns Hopkins Press, 1992.

Sahlins, Marshall. *Culture and Practical Reason.* Chicago: University of Chicago Press, 1976.

Said, Edward. *Orientalism*. New York: Vintage Press, 1994.

Sami, Amin. *Al-Ta'lim fi-l-misr fi sanatay 1914–1915*. Cairo: Matba'at al-Ma'arif, 1916.

Sayar, Z. "Bizde Mesken Finansmanı," *Arkitekt* 22, 9/10 (1952): 253–254.

———. "İnşaat Kalfaları Problemi," *Arkitekt* 17, 9/10 (1947): 199–200.

———. "I. Türk Yapı Kongresinden Beklediklerimiz," *Arkitekt* 18, 1/2 (1948): 19–34.

———. "Mesken Davası I & II," *Arkitekt* 16, 5/6 (1946): 171–172 and 16, 7/8, (1946): 149–150.

———. "1952 Mesken Faaliyeti Nasıl Olacak?" *Arkitekt* 21, 11/12 (1951): 205, 232.

Scanlon, Jennifer. *Inarticulate Longings: The Ladies Home Journal, Gender, and the Promises of Consumer Culture*. New York: Routledge, 1995.

Schneirov, Matthew. *The Dream of a New Social Order: Popular Magazines in America, 1893–1914*. New York: Columbia University Press, 1994.

Schudson, Michael. *Advertising, The Uneasy Persuasion: Its Dubious Impact on American Society*. USA: Basic Books, 1984.

Scott, James. *Seeing Like a State: How Certain Schemes to Improve on the Human Condition have Failed*. New Haven: Yale University Press, 1998.

Sela-Sheffy, Rakefet. "Models and Habituses: Problems in the Idea of Cultural Repertoires," *Canadian Review of Comparative Literature* 24, 1 (1997): 35–47.

Semyonov, Moshe, Noah Lewin-Epstein and Seymor Spilerman. "The Material Possessions of Israeli Ethnic Groups," *European Sociological Review* 12, 3 (1996): 289–301.

Şenyapılı, Tansı. "Örgütlenemeyen Nüfusa Örgütlü Çözüm: Çözümsüzlük," *Konut Araştırmaları Sempozyumu*, Konut Araştırmaları Dizisi 1. Ankara: T.C. Toplu Konut İdaresi Başkanlığı, 1995.

Seton, Grace Thompson. *A Woman Tenderfoot in Egypt*. New York: Dodd, Mead & Company, 1900.

Shafir, Gershon and Yoav Peled. *Being Israeli: The Dynamics of Multiple Citizenship*. Cambridge: Cambridge University Press, 2002.

Al-Shayyal, Jamal al-Din. *Tarikh al-tarjama wa-l-haraka al-thaqafiyya*. Cairo: Dar al-Fikr al-'Araby, 1951.

Shevelow, Katherine. *Women and Print Culture: The Construction of Femininity in the Early Periodical*. London: Routledge, 1989.

Silvera, Alain. "The First Egyptian Student Mission to France under Muhammad 'Ali." In *Modern Egypt: Studies in Politics and Society*, edited by Elie Kedourie and Sylvia Haim, 1–22. London: Frank Kass, 1980.

Silverstone, Roger, ed. *Visions of Suburbia*. London: Routledge, 1997.

Singerman, Diane. *Avenues of Participation: Family, Politics, and Networks in Urban Quarters of Cairo*. Princeton: Princeton University Press, 1995.

——— and Homa Hoodfar, eds. *Development, Change, and Gender in Cairo: A View from the Household*. Bloomington: Indiana University Press, 1996.

Sirageldin, Ismail, ed. *Human Capital: Population Economics in the Middle East*. London : I.B. Tauris, 2002.

Sislian, J.H. "Missionary Work in Egypt During the 19th Century." In *Education and the Mission Schools: Case Studies in the British Empire*, edited by Brian Holmes, New York: Humanities Press, 1967.

Slater, Don. *Consumer Culture and Modernity*. Cambridge: Polity, 1997.

——— and Fran Tonkiss. *Market Society: Markets and Modern Social Theory*. Cambridge: Polity, 2001.

Slyomovics, Susan. *The Object of Memory: Jew and Arab Narrate the Palestinian Village*. Philadelphia: University of Pennsylvania Press, 1999.

Spigel, Lynn. *Make Room for TV: Television and the Family Ideal in Postwar America*. Chicago: University of Chicago Press, 1992.

———. *Welcome to the Dreamhouse: Popular Media and Postwar Suburbs*. Durham: Duke University Press, 2001.

Steward, Desmond. *Great Cairo: Mother of the World*. Cairo: The American University in Cairo Press, 1996.

Stewart, Frances. *Planning to Meet Basic Needs*. London: Macmillan, 1985.

Tagg, John. *The Burden of Representation: Essays on Photographies and Histories*. Amherst: University of Massachusetts Press, 1988.

Al-Tahtawi, Rifa`a Rafa`i. M*anahij al-albab al-misriyya fi mabahij al-adab al-`asriyya*. 2nd ed. Cairo: Matba`at Shirkat al-Ragha'ib, 1912.

———. *Murshid al-amin li-l-banat wa-l-bani*. Cairo: 1289 AH. [c. 1872].

———. *Qala`id al-mafakhir fi gharib `awa'id al-awa'il wa-l-awakhir*. Bulaq, 1833.

———. *Takhlis al-ibriz fi-talkhis baris*. In *Al-A`mal al-kamila*. Edited by Muhammad `Imara. Beirut: Al-Mu'assa al-`Arabiyya li-l-Dirasaat wa-l-Nashr, 1973.

———. *Al-Ta`ribat al-safiyya li-murid al-jeographiyya*. Bulaq, 1834.

Tamraz, Nihal. "Nineteenth Century Domestic Architecture: Abbasia as a Case Study," M.A. thesis, American University in Cairo, 1993.

———. *Nineteenth-Century Houses and Palaces*. Cairo: The American University in Cairo Press, 1998.

Tang, Xiobing. *Global Space and the Nationalist Discourse: The Historical Thinking of Liang Qichao*. Stanford: Stanford University Press, 1996.

Tanyeli, Uğur. "Osmanlı Barınma Kültüründe Batılılaşma-Modernleşme: Yeni Bir Simgeler Dizisinin Oluşumu." In *Housing and Settlement in Anatolia in Historical Perspective*, edited by Yıldız Sey, 284–297. Istanbul: History Foundation Publications, 1996.

Tekeli, İlhan. "70 Yıl İçinde Türkiye'de Konut Sorununa Nasıl Çözüm Arandı?" *Konut Araştırmaları Sempozyumu, Konut Araştırmaları Dizisi I*, 1–10, Ankara: Toplu Konut İdaresi Başkanlığı, 1995.

The State Institute of Statistics, *Statistical Yearbook*. Various issues. (Turkish).

Tignor, Robert. "Bank Misr and Foreign Capitalism." *International Journal of Middle East Studies* 8 (1977): 161–181.

Toledano, Ehud. "*Social and Economic Change in 'The Long Nineteenth Century'*." In *The Cambridge History of Egypt*, edited by M.W. Daly. Cambridge: Cambridge University Press, 1998.

——. *State and Society in Mid-Nineteenth-Century Egypt*. Cambridge: Cambridge University Press, 1990.

Toury, Gideon. "Culture Planning and Translation." Forthcoming in *Proceedings of the Vigo Conference "anovadores de nós–anosadores de vós,"* edited by Alberto Alvarez et al.

Troutt-Powell, Eve M. "From Odyssey to Empire: Mapping Sudan through Egyptian Literature in the Mid-Nineteenth Century," *International Journal of Middle East Studies* 31 (1999): 401–427.

Tsevet Tesh`a le-Tichnun Kollel, 1969. *Julis, Deir al-Asad, Nahf: Economic and Social Survey*. Tel Aviv, Segal ba-`am Press, 1969 (Hebrew).

Veblen, Thorstein. *The Theory of the Leisure Class*. New York: Viking Press, 1931 [1899].

Walker, Nancy. *Shaping Our Mother's World: American Women's Magazines*. Mississippi: University of Mississippi Press, 2000.

Watson, Andrew. *The American Mission in Egypt, 1854–1896*. Pittsburgh, 1898.

Watson, Charles. *Egypt and the Christian Crusade*. Philadelphia: United Presbyterian Church of North American, 1907.

Wigh, Leif. *Photographic Views of the Bosphorus and Constantinople*. Stockholm: Fotografiska Museet, 1984. (Swedish).

Wikan, Unni. "Living Conditions amongst Cairo's Poor," *Middle East Journal* 35, 1 (1985): 7–26.

——. *Tomorrow, God Willing: Self-made Destinies in Cairo*. Chicago: University of Chicago Press, 1996.

Wilson, Lara. *The Photo Album of Nellie L. McClung (1873–1951)*. M.A. thesis, University of Victoria, 1998.

Wilkinson, Sir I. Gardner. *A Handbook for Travellers to Egypt*. London, 1867.

Williams, Rosalind. *Dream Worlds: Mass Consumption in Late-Nineteenth-Century France*. Berkeley: University of California Press, 1983.

Winichakul, Thongchai. *Siam Mapped: A History of the Geo-Body of a Nation*. Honolulu: University of Hawaii Press, 1992.

Winship, Janice. *Inside Women's Magazines*. London: Pandora Press, 1987.

Wolfe, Patrick. "History of Imperialism: A Century of Theory, from Marx to Postcolonialism," *The American Historical Review* 102, 2 (1997): 388–420.

Yiftachel, Oren. *Guarding the Grove*. Beit Berl: The Institute for Israeli Arab Studies, 1997. (Hebrew).

——. *Planning a Mixed Region in Israel: The Political Geography of Arab–Jewish Relations in the Galilee*. London: Avebury-Gower, 1992.

Zelizer, Viviana. *The Social Meanings of Money*. Princeton: Princeton University Press, 1997.

Ziya, Abdullah. "Binanın içinde Mimar," *Mimar* 1, 1 (1931): 13–17.

Index

DATE DUE

Printed
in USA

HIGHSMITH #45230